AS Maths
Edexcel Core 2

AS level maths is seriously tricky — no question about that.

We've done everything we can to make things easier for you.
We've scrutinised past paper questions and we've gone through the syllabuses with a fine-toothed comb. So we've found out exactly what you need to know, then explained it simply and clearly.

We've stuck in as many helpful hints as you could possibly want
— then we even tried to put some funny bits in to keep you awake.

We've done our bit — the rest is up to you.

Contents

SECTION 6 — INTEGRATION

P1 — PRACTICE EXAM 1

P1 — PRACTICE EXAM 2

This book covers the Core 2 module of the Edexcel specification.

Published by Coordination Group Publications Ltd.

Contributors:
Charley Darbishire
Simon Little
Andy Park
Glenn Rogers
Claire Thompson

And:
Iain Nash

Updated by:
Sharon Keeley
Tim Major
Sam Norman
Andy Park
Alan Rix
Claire Thompson
Julie Wakeling

ISBN: 1-84146-768-5

Groovy website: www.cgpbooks.co.uk

Jolly bits of clipart from CorelDRAW
With thanks to Colin Wells and Janet Dickinson for the proofreading.

Printed by Elanders Hindson, Newcastle upon Tyne.

Algebraic Division

Algebraic division is one of those things that you have to learn when you do AS maths.
You'll probably never use it again once you've done your Exam, but hey ho... such is life.

Do **Algebraic Division** by means of **Subtraction**

$$(2x^3 - 3x^2 - 3x + 7) \div (x - 2) = ?$$

The trick with this is to see how many times you can <u>subtract</u> $(x-2)$ from $2x^3 - 3x^2 - 3x + 7$.
The idea is to keep <u>subtracting</u> lumps of $(x-2)$ until you've got rid of all the <u>powers of x</u>.

Do the subtracting in **Stages**

At each stage, always try to get rid of the <u>highest</u> power of x.
Then start again with whatever you've got left.

① Start with $2x^3 - 3x^2 - 3x + 7$, and <u>subtract</u> $2x^2$ lots of $(x-2)$ to get rid of the x^3 term.

$$(2x^3 - 3x^2 - 3x + 7) - 2x^2(x-2)$$
$$(2x^3 - 3x^2 - 3x + 7) - 2x^3 + 4x^2$$
$$= x^2 - 3x + 7$$

> $2x^3 \div x = 2x^2$

> This is what's left — so now you have to get rid of the $\underline{x^2}$ term.

② Now <u>start again</u> with $x^2 - 3x + 7$.
The highest power of x is the x^2 term.
So <u>subtract</u> x lots of $(x-2)$ to get rid of that.

$$(x^2 - 3x + 7) - x(x-2)$$
$$(x^2 - 3x + 7) - x^2 + 2x$$
$$= -x + 7$$

> Now start again with this — and get rid of the $\underline{x}$ term.

③ All that's left now is $-x + 7$.
Get rid of the $-x$ by <u>subtracting</u> -1 times $(x-2)$.

$$(-x + 7) - (-1(x-2))$$
$$(-x + 7) + x - 2$$
$$= 5$$

> There are no more powers of x to get rid of — so <u>stop here</u>.

> The <u>remainder's</u> 5.

Interpreting the results...

Time to work out exactly what all that <u>meant</u>...

Started with: $2x^3 - 3x^2 - 3x + 7$

Subtracted: $2x^2(x-2) + x(x-2) - 1(x-2)$

$$= (x-2)(2x^2 + x - 1)$$

Remainder: $= 5$

So... $2x^3 - 3x^2 - 3x + 7 = (x-2)(2x^2 + x - 1) + 5$

...or to put that another way...

$$\frac{2x^3 - 3x^2 - 3x + 7}{(x-2)} = 2x^2 + x - 1 \text{ with remainder 5.}$$

> $2x^2 + x - 1$ is called the <u>quotient</u>.

Algebraic Division

$$(ax^3 + bx^2 + cx + d) \div (x - k) = ?$$

1) <u>SUBTRACT</u> a multiple of $(x - k)$ to get rid of the highest power of x.

2) <u>REPEAT</u> step 1 until you've got rid of all the powers of x.

3) <u>WORK OUT</u> how many lumps of $(x - k)$, you've subtracted, and the <u>REMAINDER</u>.

Algebraic division is a beautiful thing that we should all cherish...

Revising algebraic division isn't the most enjoyable way to spend an afternoon, it's true, but it's in the specification, and so you need to be comfortable with it. It involves the same process you use when you're doing long division with numbers — so if you're having trouble following the above, do $4863 \div 7$ really slowly. What you're doing at each stage is subtracting multiples of 7, and you do this until you can't take any more 7s away, which is when you get your remainder.

The Remainder and Factor Theorems

The Remainder Theorem and the Factor Theorem are easy, and possibly quite useful.

The **Remainder Theorem** is an easy way to work out **Remainders**

$$\text{When you divide } f(x) \text{ by } (x - a), \text{ the remainder is } f(a).$$

So in the example on the previous page, you could have worked out the remainder dead easily.

1) $f(x) = 2x^3 - 3x^2 - 3x + 7$.
2) You're dividing by $(x - 2)$, so $a = 2$.
3) So the remainder must be $f(2) = (2 \times 8) - (3 \times 4) - (3 \times 2) + 7 = 5$.

Careful now... when you're dividing by something like $(x + 7)$, a is negative — so here, a = –7.

It's no harder if you want the remainder after dividing by something like $(ax - b)$.

If you multiply the first factor by 4, you have to divide the second one by 4... ...but the remainder is unaffected.

Example: Find the remainder when you divide $2x^3 - 3x^2 - 3x + 7$ by $4x - 8$.

You know that $2x^3 - 3x^2 - 3x + 7 = (x - 2)(2x^2 + x - 1) + 5$, and so $2x^3 - 3x^2 - 3x + 7 = (4x - 8)\left[\frac{1}{2}x^2 + \frac{1}{4}x - \frac{1}{4}\right] + 5$

This means that the remainder when you divide by $4x - 8$ $(= 4(x - 2))$ is just 5 — the same as when you divide by $x - 2$.

The **Factor Theorem** is just the Remainder Theorem with a **Zero Remainder**

If you get a remainder of zero when you divide f(x) by (x – a), then (x – a) must be a factor. That's the Factor Theorem.

$$\text{The Factor Theorem:}$$
$$\text{If } f(x) \text{ is a polynomial, and } f(a) = 0, \text{ then } (x - a) \text{ is a factor of } f(x).$$
In other words: If you know the roots, you also know the factors — and vice versa.

Example: Show that $(2x + 1)$ is a factor of $f(x) = 2x^3 - 3x^2 + 4x + 3$

The question's giving you a big hint here. Notice that $2x + 1 = 0$ when $x = -\frac{1}{2}$. So plug this value of x into $f(x)$. If you show that $f(-\frac{1}{2}) = 0$, then the factor theorem says that $(x + \frac{1}{2})$ is a factor — which means that $2 \times (x + \frac{1}{2}) = (2x + 1)$ is also a factor.

$$f(x) = 2x^3 - 3x^2 + 4x + 3 \quad \text{and so} \quad f\left(-\frac{1}{2}\right) = 2 \times \left(-\frac{1}{8}\right) - 3 \times \frac{1}{4} + 4 \times \left(-\frac{1}{2}\right) + 3 = 0$$

So, by the factor theorem, $(x + \frac{1}{2})$ is a factor of $f(x)$, and so $(2x + 1)$ is also a factor.

(x – 1) is a Factor if the coefficients **Add Up To 0**

This is a useful thing to remember.
It works for all polynomials — no exceptions.
It could save a fair whack of time in the exam.

Example: Factorise the polynomial $f(x) = 6x^2 - 7x + 1$

The coefficients (6, –7 and 1) add up to 0. That means $f(1) = 0$.
(And that applies to any polynomial at all... always.)

So by the factor theorem, if $f(1) = 0$, $(x - 1)$ is a factor. Easy.
Then just factorise it like any quadratic to get this:

$$f(x) = 6x^2 - 7x + 1 = (6x - 1)(x - 1)$$

Section One Revision Questions

What a start that was. One thing that's definitely true about that opening section is that it was short. It was 2 pages long, to be precise. That's about as short as it could be without being really silly. So stop complaining that you're hard done by and have a crack at these questions. There aren't many of them, because you haven't done much yet. But if you get something wrong, look at the last couple of pages again to see where you went astray, then try again. You should be able to get them all right. So if you can't, you've got some more learning to do.

1) Write the following functions $f(x)$ in the form $f(x) = (x + 2)g(x) + \text{remainder}$ (where $g(x)$ is a quadratic):
   a) $f(x) = 3x^3 - 4x^2 - 5x - 6$,
   b) $f(x) = x^3 + 2x^2 - 3x + 4$

2) Find the remainder when the following are divided by: (i) $(x + 1)$,   (ii) $(x - 1)$
   a) $f(x) = 6x^3 - x^2 - 3x - 12$,
   b) $f(x) = x^4 + 2x^3 - x^2 + 3x + 4$

3) Find the remainder when $f(x) = x^4 - 3x^3 + 7x^2 - 12x + 14$ is divided by:
   a) $x + 2$
   b) $2x + 4$

4) Which of the following are factors of $f(x) = x^5 - 4x^4 + 3x^3 + 2x^2 - 2$?
   a) $x - 1$
   b) $x + 1$
   c) $x - 2$

5) Find the values of c and d so that $2x^4 + 3x^3 + 5x^2 + cx + d$ is exactly divisible by $(x - 2)(x + 3)$.

Circles

I always say a beautiful shape deserves a beautiful formula, and here you've got one of my favourite double-acts...

Equation of a circle: $(x - a)^2 + (y - b)^2 = r^2$

The equation of a circle looks complicated, but it's all based on Pythagoras' theorem.
Take a look at the circle below, with centre (6, 4) and radius 3.

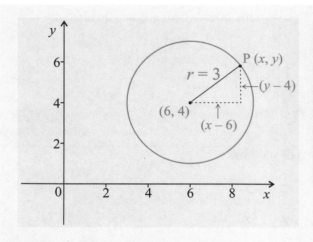

Joining a point P (x, y) on the circumference of the circle to its centre (6, 4), we can create a right-angled triangle.

Now let's see what happens if we use Pythagoras' theorem:

$$(x - 6)^2 + (y - 4)^2 = 3^2$$

or: $(x - 6)^2 + (y - 4)^2 = 9$

This is the equation for the circle. It's as easy as that.

In general, a circle with radius r and centre (a, b) has the equation: $(x - a)^2 + (y - b)^2 = r^2$

Example:

i)   What is the centre and radius of the circle with equation $(x - 2)^2 + (y + 3)^2 = 16$

ii) Write down the equation of the circle with centre (–4, 2) and radius 6.

Solution:

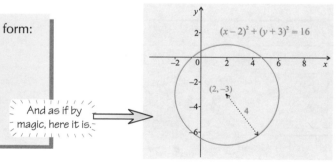

i)   Comparing $(x - 2)^2 + (y + 3)^2 = 16$ with the general form:

$$(x - a)^2 + (y - b)^2 = r^2$$

then $a = 2$, $b = -3$ and $r = 4$.

So the centre (a, b) is: (2, –3)
and the radius (r) is: 4.

And as if by magic, here it is.

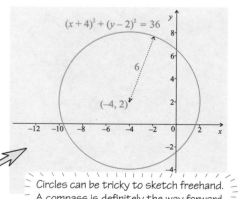

ii) The question says, 'Write down...', so you know you don't need to do any working.
The centre of the circle is (–4, 2), so $a = -4$ and $b = 2$.
The radius is 6, so $r = 6$.
Using the general equation for a circle $(x - a)^2 + (y - b)^2 = r^2$
you can write: $(x + 4)^2 + (y - 2)^2 = 36$

Circles can be tricky to sketch freehand. A compass is definitely the way forward.

This is pretty much all you need to learn. Everything on the next page uses stuff you should know already.

Circles

Rearrange the equation into the *familiar form*

Sometimes you'll be given an equation for a circle that doesn't look much like $(x - a)^2 + (y - b)^2 = r^2$.
This is a bit of a pain, because it means you can't immediately tell what the **radius** is or where the **centre** is.
But all it takes is a bit of **rearranging**.

Let's take the equation: $x^2 + y^2 - 6x + 4y + 4 = 0$

You need to get it into the form $(x - a)^2 + (y - b)^2 = r^2$

This is just like completing the square.

> Have a look at Core 1 pages 12-13
> for more on completing the square.

$x^2 + y^2 - 6x + 4y + 4 = 0$
$x^2 - 6x + y^2 + 4y + 4 = 0$
$(x - 3)^2 - 9 + (y + 2)^2 - 4 + 4 = 0$
$(x - 3)^2 + (y + 2)^2 = 9 \Longrightarrow$ This is the recognisable form, so the centre is **(3, –2)** and the radius is $\sqrt{9} = 3$.

Don't forget the *Properties of Circles*

You will have seen the circle rules at GCSE. You'll sometimes need to dredge them up in your memory
for these circle questions. Here's a reminder of a few useful ones.

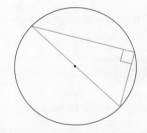

The angle in a semicircle is a right angle.

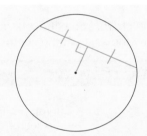

The perpendicular from the centre to a chord bisects the chord.

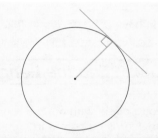

A radius and tangent to the same point will meet at right angles.

Use the *Gradient Rule* for Perpendicular Lines

Remember that the tangent at a given point will be perpendicular to the radius at that same point.

Example:

Point A (6, 4) lies on a circle with the equation $x^2 + y^2 - 4x - 2y - 20 = 0$.
  i) Find the centre and radius of the circle.
  ii) Find the equation of the tangent to the circle at A.

Solution:

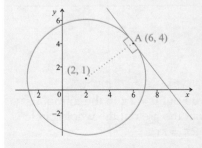

i) Rearrange the equation to show it as the sum of 2 squares:
$x^2 + y^2 - 4x - 2y - 20 = 0$
$x^2 - 4x + y^2 - 2y - 20 = 0$
$(x - 2)^2 - 4 + (y - 1)^2 - 1 - 20 = 0$
$(x - 2)^2 + (y - 1)^2 = 25$
This shows the centre is **(2, 1)** and the radius is **5**.

ii) The tangent is at right angles to the radius at (6, 4).

Gradient of radius at (6, 4) $= \dfrac{4 - 1}{6 - 2} = \dfrac{3}{4}$

Gradient of tangent $= \dfrac{-1}{\frac{3}{4}} = -\dfrac{4}{3}$

Using $y - y_1 = m(x - x_1)$
$y - 4 = -\dfrac{4}{3}(x - 6)$
$3y - 12 = -4x + 24$
$3y + 4x - 36 = 0$

So the chicken comes from the egg, and the egg comes from the chicken...

Well folks, at least it makes a change from all those straight lines and quadratics.
I reckon if you know the **formula** and **what it means**, you should be absolutely **fine** with questions on circles.

Arc Length and Sector Area

Arc lengths and sector areas are easier than you'd think — once you've learnt two simple(ish) formulas.

Always work in *Radians* for *Arc Length* and *Sector Area Questions*

Remember — for arc length and sector area questions you've got to measure all the angles in <u>radians</u>.
The main thing is that you know how radians relate to <u>degrees</u>.
In short, 180 degrees = π radians. The table below shows you how to convert between the two units:

Converting angles	
<u>Radians to degrees:</u>	<u>Degrees to radians:</u>
Divide by π, multiply by 180.	**Divide by 180, multiply by π.**

Here's a table of some of the common angles you're going to need — in degrees and radians:

Degrees	0	30	45	60	90	120	180	270	360
Radians	0	$\dfrac{\pi}{6}$	$\dfrac{\pi}{4}$	$\dfrac{\pi}{3}$	$\dfrac{\pi}{2}$	$\dfrac{2\pi}{3}$	π	$\dfrac{3\pi}{2}$	2π

If you have <u>part of a circle</u> (like a section of pie chart), you can work out the <u>length of the curved side</u>, or the <u>area of the 'slice of pie'</u> — as long as you know the <u>angle</u> at the centre (θ) and the <u>length of the radius</u> (r). Read on...

You can find the *Length* of an *Arc* using a nice easy formula...

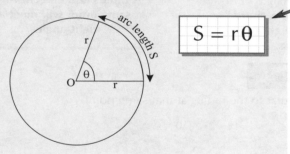

For a circle with a <u>radius of r</u>, where the angle θ is measured in <u>radians</u>, the <u>arc length of the sector S</u> is given by:

$$S = r\theta$$

If you put θ = 2π in this formula (and so make the sector equal to the whole circle), you get that the distance all the way round the outside of the circle is S = 2πr.

This is just the normal circumference formula.

...and the area of a *Sector* using a similar formula

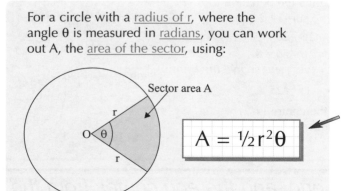

For a circle with a <u>radius of r</u>, where the angle θ is measured in <u>radians</u>, you can work out A, the <u>area of the sector</u>, using:

Sector area A

$$A = \tfrac{1}{2}r^2\theta$$

Again, if you put θ = 2π in the formula, you find that the area of the whole circle is A = ½r² × 2π = πr².

This is just the normal 'area of a circle' formula.

Arc Length and Sector Area

Questions on <u>trigonometry</u> quite often use the same angles — so it makes life easier if you know the sin, cos and tan of these commonly used angles. Or to put it another way, examiners expect you to know them — so learn them.

Draw Triangles to remember *sin, cos* and *tan* of the *Important Angles*

You should know the values of <u>sin</u>, <u>cos</u> and <u>tan</u> at 30°, 60° and 45°. But to help you remember, you can draw these two groovy triangles. It may seem a complicated way to learn a few numbers, but it does make it easier. Honest.

The idea is you draw the triangles below, putting in their angles and side lengths. Then you can use them to work out special trig values like <u>sin 45°</u> or <u>cos 60°</u> more accurately than any calculator (which only gives a few decimal places).

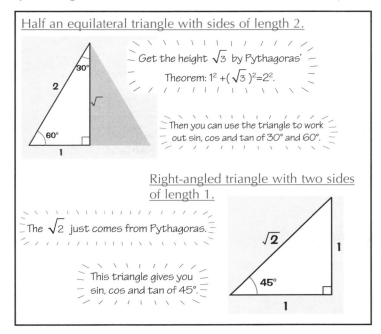

Half an equilateral triangle with sides of length 2.

Get the height $\sqrt{3}$ by Pythagoras' Theorem: $1^2 + (\sqrt{3})^2 = 2^2$.

Then you can use the triangle to work out sin, cos and tan of 30° and 60°.

Right-angled triangle with two sides of length 1.

The $\sqrt{2}$ just comes from Pythagoras.

This triangle gives you sin, cos and tan of 45°.

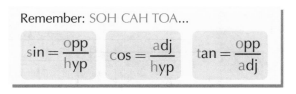

Remember: SOH CAH TOA...

$$\sin = \frac{\text{opp}}{\text{hyp}} \qquad \cos = \frac{\text{adj}}{\text{hyp}} \qquad \tan = \frac{\text{opp}}{\text{adj}}$$

Trig Values from Triangles

$$\sin 30° = \frac{1}{2} \qquad \sin 60° = \frac{\sqrt{3}}{2} \qquad \sin 45° = \frac{1}{\sqrt{2}}$$

$$\cos 30° = \frac{\sqrt{3}}{2} \qquad \cos 60° = \frac{1}{2} \qquad \cos 45° = \frac{1}{\sqrt{2}}$$

$$\tan 30° = \frac{1}{\sqrt{3}} \qquad \tan 60° = \sqrt{3} \qquad \tan 45° = 1$$

Example: Find the exact length L and area A in the diagram.

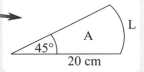

Right, first things first... it's an arc length and sector area, so you need the angle in radians.

$$45° = \frac{45 \times \pi}{180} = \frac{\pi}{4} \text{ radians}$$

Or you could just quote this if you've learnt the stuff above.

Now bung everything in your formulas:

$$L = r\theta = 20 \times \frac{\pi}{4} = 5\pi \text{ cm}$$

$$A = \frac{1}{2}r^2\theta = \frac{1}{2} \times 20^2 \times \frac{\pi}{4} = 50\pi \text{ cm}^2$$

Example: Find the area of the shaded part of the symbol.

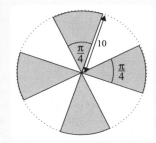

You need the area of the 'leaves' and so use the formula ½r²θ.

Each leaf has area $\frac{1}{2} \times 10^2 \times \frac{\pi}{4} = 25\frac{\pi}{2} \text{ cm}^2$

So the area of the whole symbol $= 4 \times 25\frac{\pi}{2} = 50\pi \text{ cm}^2$

π = 3.141592653589793238462643383279502884197169399...*(Make sure you know it)*

It's worth repeating, just to make sure — those formulas for arc length and sector area only work if the angle is in <u>radians</u>.

The Trig Formulas You Need to Know

There are some more trig formulas you need to know for the exam.
So here they are — learn them or you're seriously stuffed. Worse than an aubergine.

*The **Sine Rule** and **Cosine Rule** work for **Any** triangle*

Remember these three formulas work for <u>ANY</u> triangle, not just right-angled ones.

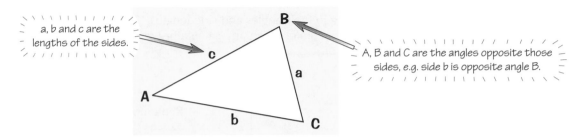

> a, b and c are the lengths of the sides.

> A, B and C are the angles opposite those sides, e.g. side b is opposite angle B.

THE SINE RULE
$$\frac{a}{\sin A} = \frac{b}{\sin B} = \frac{c}{\sin C}$$

THE COSINE RULE
$$a^2 = b^2 + c^2 - 2bc\cos A$$

AREA OF ANY TRIANGLE
$$Area = \tfrac{1}{2}ab\sin C$$

Sine Rule or Cosine Rule — which one is it...

To decide which of these two rules you need to use, look at how much you <u>already</u> know about the triangle.

<u>Sine Rule</u>

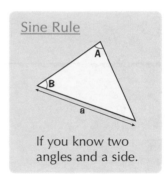

If you know two angles and a side.

<u>Cosine Rule — Then Sine Rule</u>

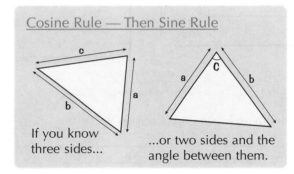

If you know three sides...

...or two sides and the angle between them.

<u>Err — it doesn't work here...</u>

If you've got two sides and an angle (but not the angle between them)...

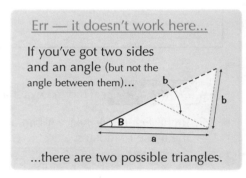

...there are two possible triangles.

*The **Best** has been saved till last...*

These two identities are really important. You'll need them <u>loads</u>.

$$\tan x \equiv \frac{\sin x}{\cos x}$$

$$\sin^2 x + \cos^2 x \equiv 1$$

$$\Rightarrow \sin^2 x \equiv 1 - \cos^2 x$$
$$\cos^2 x \equiv 1 - \sin^2 x$$

> Work out these two using sin²x + cos²x ≡ 1.

These two come up in exam questions <u>all the time</u>. Learn them.
Learnthemlearnthemlearnthemlearnthemlearnthemlear... okay, I'll stop now.

Tri angles — go on... you might like them.

Formulas and trigonometry go together even better than Richard and Judy. I can count 7 on this page. That's not many, so please, just make sure you know them! If you haven't learned them I will cry for you. I will sob.

Using the Sine and Cosine Rules

This page is about "solving" triangles, which just means finding all their sides and angles when you already know a few.

EXAMPLE: Solve $\triangle$ ABC, in which A=40°, a=27 m, B=73°. Then find the area.

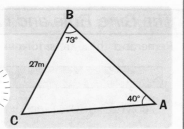

Draw a quick sketch first — don't worry if it's not deadly accurate, though.
You're given 2 angles and a side, so you need the Sine Rule.

Make sure you put side a opposite angle A.

First of all, get the other angle: $\angle C = (180 - 40 - 73)° = 67°$

Then find the other sides, one at a time:

$$\frac{a}{\sin A} = \frac{b}{\sin B} \Rightarrow \frac{27}{\sin 40°} = \frac{b}{\sin 73°}$$
$$\Rightarrow b = \frac{\sin 73°}{\sin 40°} \times 27 = \underline{40.2\,m}$$

$$\frac{c}{\sin C} = \frac{a}{\sin A} \Rightarrow \frac{c}{\sin 67°} = \frac{27}{\sin 40°}$$
$$\Rightarrow c = \frac{\sin 67°}{\sin 40°} \times 27 = \underline{38.7\,m}$$

Now just use the formula to find its area.

$$\text{Area of } \triangle \text{ ABC} = \tfrac{1}{2} ab \sin C$$
$$= \tfrac{1}{2} \times 27 \times 40.169 \times \sin 67°$$
$$= \underline{499.2\,m^2}$$

Use a more accurate value for b here, rather than the rounded value 40.2.

EXAMPLE: Find X, Y and z.

You've been given 2 sides and the angle between them, so you're going to need the Cosine Rule, then the Sine Rule.

$$\boxed{a^2 = b^2 + c^2 - 2bc \cos A}$$ $$\boxed{\frac{a}{\sin A} = \frac{b}{\sin B} = \frac{c}{\sin C}}$$

$$z^2 = (6.5)^2 + 10^2 - 2(6.5)(10) \cos 35°$$
$$\Rightarrow z^2 = 142.25 - 130 \cos 35°$$
$$\Rightarrow z^2 = 35.7602$$
$$\Rightarrow z = 5.98\,cm$$

In this case, angle A is 35°, and side a is actually z.

You've got all the sides. Now use the Sine Rule to find the other two angles.

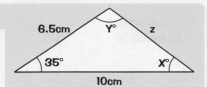

$$\frac{6.5}{\sin X} = \frac{5.9800}{\sin 35°}$$
$$\Rightarrow \sin X = 0.6235$$
$$\Rightarrow X = \sin^{-1} 0.6235$$
$$\Rightarrow \underline{X = 38.6°}$$

$$\frac{10}{\sin Y} = \frac{5.9800}{\sin 35°}$$
$$\Rightarrow \sin Y = 0.9592$$
$$\Rightarrow Y = \sin^{-1} 0.9592$$
$$\Rightarrow Y = 73.6° \text{ or } \underline{106.4°}$$

*The Sine Rule just says:
Dividing "length of a side" by "the sine of the opposite angle" gives the same answer, whichever pair you choose.*

*This is the answer you need.
Be careful: your calculator only gives you values for sin⁻¹ between −90° and 90°.*

Check your answers by adding up all the angles in the triangle. If they don't add up to 180°, you've gone wrong somewhere.

Graphs of Trig Functions

Before you leave this page, you should be able to close your eyes and picture these three graphs in your head, properly labelled and everything. If you can't, you need to learn them more. I'm not kidding.

sin x and cos x are always in the range –1 to 1

sin x and cos x are similar — they just bob up and down between –1 and 1.

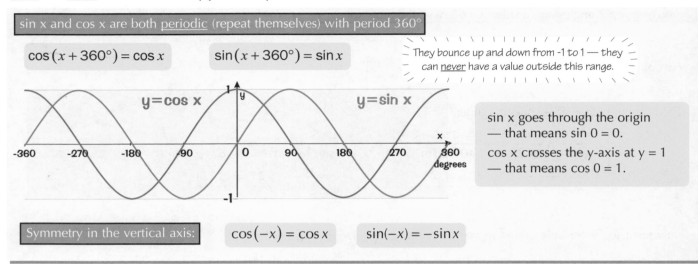

sin x and cos x are both <u>periodic</u> (repeat themselves) with period 360°

$$\cos(x + 360°) = \cos x \qquad \sin(x + 360°) = \sin x$$

They bounce up and down from -1 to 1 — they can <u>never</u> have a value outside this range.

sin x goes through the origin — that means sin 0 = 0.

cos x crosses the y-axis at y = 1 — that means cos 0 = 1.

Symmetry in the vertical axis: $\cos(-x) = \cos x \qquad \sin(-x) = -\sin x$

tan x can be Any Value at all

tan x is different from sin x or cos x.
It doesn't go gently up and down between –1 and 1 — it goes between $-\infty$ and $+\infty$.

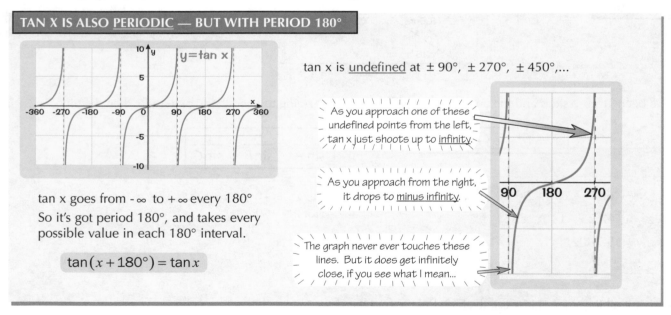

TAN X IS ALSO <u>PERIODIC</u> — BUT WITH PERIOD 180°

tan x is <u>undefined</u> at $\pm 90°$, $\pm 270°$, $\pm 450°$,...

As you approach one of these undefined points from the left, tan x just shoots up to <u>infinity</u>.

As you approach from the right, it drops to <u>minus infinity</u>.

tan x goes from $-\infty$ to $+\infty$ every 180°
So it's got period 180°, and takes every possible value in each 180° interval.

$$\tan(x + 180°) = \tan x$$

The graph never ever touches these lines. But it does get infinitely close, if you see what I mean...

The easiest way to sketch any of these graphs is to plot the important points which happen every 90° (i.e. –180°, –90°, 0°, 90°, 180°, 270°, 360°...) and then just join the dots up.

Sin and cos can make your life worthwhile — give them a chance...

It's really really really really really important that you can draw the trig graphs on this page, and get all the labels right. Make sure you know what value sin, cos and tan have at the interesting points — i.e. 0°, 90°, 180°, 270°, 360°. It's easy to remember what the graphs look like, but you've got to know exactly <u>where</u> they're max, min, zero, etc.

Transformed Trig Graphs

Transformed trigonometric graphs look much the same as the bog standard ones, just a little different.
There are three main types of transformation.

There are 3 basic types of Transformed Trig Graph...

$y = n\sin x$ — a Vertical Stretch or Squash

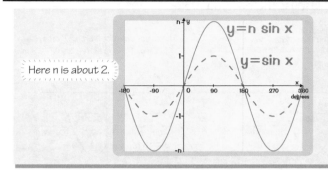

Here n is about 2.

If n > 1, the graph of y = sin x is
stretched vertically by a factor of n.

If 0 < n < 1, the graph is squashed.

And if n < 0, the graph is also
reflected in the x-axis.

$y = \sin nx$ — a Horizontal Squash or Stretch

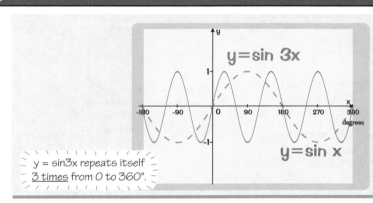

y = sin3x repeats itself
3 times from 0 to 360°.

If n > 1, the graph of y = sin x is
squashed horizontally by a factor of n.

If 0 < n < 1, the graph is stretched.

And if n < 0, the graph is also
reflected in the y-axis.

$y = \sin(x + c)$ — a Translation along the x-axis

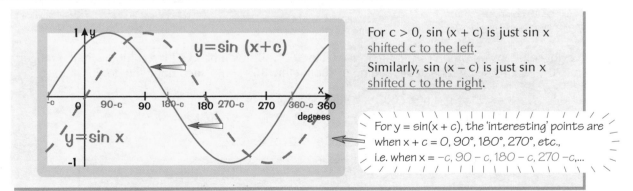

For c > 0, sin (x + c) is just sin x
shifted c to the left.
Similarly, sin (x – c) is just sin x
shifted c to the right.

For y = sin(x + c), the 'interesting' points are
when x + c = 0, 90°, 180°, 270°, etc.,
i.e. when x = –c, 90 – c, 180 – c, 270 –c,...

Curling up on the sofa with 2cos x — that's my idea of cosiness ☺

One thing you've really got to be careful about is making sure you move or stretch the graphs in the right direction. In that last example, the graph would have moved to the right if "c" was negative. And it gets confusing with the horizontal and vertical stretching and squashing — in the first two examples above, n > 1 means a vertical stretch but a horizontal squash.

SECTION TWO — CIRCLES AND TRIGONOMETRY

Solving Trig Equations in a Given Interval

I used to really hate trig stuff like this. But once I'd got the hang of it, I just couldn't get enough. I stopped going out, lost interest in the opposite sex — the CAST method became my life. Learn it, but be careful. It's addictive.

There are **Two Ways** to find Solutions in an **Interval**...

EXAMPLE: Solve cos x = $\frac{1}{2}$ for -360° ≤ x ≤ 720°.

Like I said — there are two ways to solve this kind of question. Just use the one you prefer...

You can draw a graph...

Your calculator gives you a solution of 60° (but see page 7 if you didn't know this anyway). Then you have to work out what the others will be.

The other solutions are 60° either side of the graph's peaks.

1) Draw the graph of y = cosx for the range you're interested in...

2) Get the first solution from your calculator and mark this on the graph,

3) Use the symmetry of the graph to work out what the other solutions are:

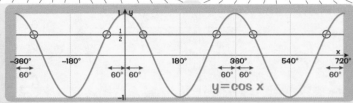

So the solutions are: −300°, −60°, 60°, 300°, 420° and 660°.

...or you can use the **CAST** diagram

<u>CAST</u> stands for <u>COS</u>, <u>ALL</u>, <u>SIN</u>, <u>TAN</u> — and the CAST diagram shows you where these functions are <u>positive</u>:

Between 90° and 180°, only SIN is positive.

Between 0 and 90°, ALL of sin, cos and tan are positive.

90°

S A

180°   x°   0°

T C

270°

Between 180° and 270°, only TAN is positive.

Between 270° and 360°, only COS is positive.

This is positive — so you're only interested in where cos is positive.

First, to find all the values of x between 0° and 360° where cos x = $\frac{1}{2}$ — you do this:

Put the first solution onto the CAST diagram.	Find the other angles between 0° and 360° that might be solutions.	Ditch the ones that are the wrong sign.

The angle from your calculator goes <u>anticlockwise</u> from the x-axis (unless it's negative — then it would go clockwise into the 4th quadrant).

The other solutions come from making the <u>same angle from the horizontal</u> axis into the other 3 quadrants.

cos x = ½, which is <u>positive</u>. The CAST diagram tells you cos is positive in the 4th quadrant — but not the 2nd or 3rd — so ditch those two angles.

So you've got solutions 60° and 300° in the range 0° to 360°. But you need all the solutions in the range −360° to 720°. Get these by repeatedly adding or subtracting 360° onto each until you go out of range:

x = 60° ⇒ (<u>adding</u> 360°) x = 420°, 780° (too big)

and (<u>subtracting</u> 360°) x = −300°, −660° (too small)

x = 300° ⇒ (<u>adding</u> 360°) x = 660°, 1020° (too big)

and (<u>subtracting</u> 360°) x = −60, −420° (too small)

So the solutions are: x = −300°, −60°, 60°, 300°, 420° and 660°.

And I feel that love is dead, I'm loving angles instead...

Suppose the first solution you get is <u>negative</u>, let's say –d°, then you'd measure it <u>clockwise</u> on the CAST diagram. So it'd be d° in the 4th quadrant. Then you'd work out the other 3 possible solutions in exactly the same way, rejecting the ones which weren't the right sign. Got that? No? Got that? No? Got that? Yes? Good!

Solving Trig Equations in a Given Interval

Sometimes it's a bit more complicated. But only a bit.

Sometimes you end up with *sin kx = number*...

For these, it's definitely easier to draw the graph rather than use the CAST method — that's one reason why being able to sketch these trig graphs properly is so important.

EXAMPLE: Solve: $\sin 3x = -\frac{1}{\sqrt{2}}$ for $0° \leq x \leq 360°$.

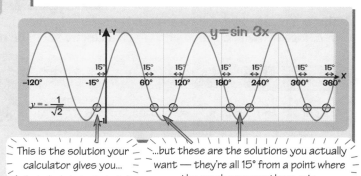

1) You've got 3x instead of x — so when you draw the graph, make it three times as <u>squashed</u> (the period will be 120° instead of 360°).

2) When you use your calculator to get the first solution, it'll probably give you a <u>negative</u> answer (which you don't want).

$$\sin 3x = -\frac{1}{\sqrt{2}}$$
$$\Rightarrow 3x = -45°$$
$$\Rightarrow x = -15°$$

You don't want this solution — so use your graph to work out the ones that you do want.

> This is the solution your calculator gives you...
>
> ...but these are the solutions you actually want — they're all 15° from a point where the graph crosses the x-axis.

3) Using the <u>symmetry</u> of the graph, you can see that the solutions you want are:

$$x = 75°, 105°, 195°, 225°, 315° \text{ and } 345°.$$

> It really is mega-important that you check these answers — it's dead easy to make a silly mistake.

4) <u>Check</u> your answers by putting these values back into your calculator.

...or *sin (x + k) = number*

All the steps in this example are just the same as in the one above.

EXAMPLE: Solve $\sin(x + 60°) = \frac{3}{4}$ for $-360° \leq x \leq 360°$, giving your answers to 2 decimal places.

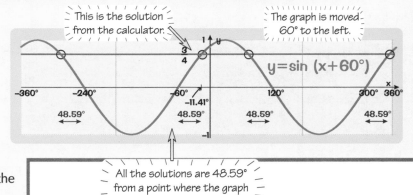

1) You've got sin (x + 60°) instead of sin x — so when you draw the graph, you have to move it 60° to the <u>left</u>.

2) Use your calculator to get that first solution...

$$\sin(x + 60°) = \frac{3}{4}$$
$$\Rightarrow x + 60° = 48.59°$$
$$\Rightarrow x = -11.41°$$

> This is the solution from the calculator.
>
> The graph is moved 60° to the left.
>
> All the solutions are 48.59° from a point where the graph crosses the x-axis.

3) And again, use the graph's <u>symmetry</u> to get the other three. The solutions are:

$$x = -11.41°, -288.59°, 71.41° \text{ and } 348.59°$$

4) <u>Check</u> your answers by putting these values back into your calculator.

Live a life of sin (and cos and tan)...

Yep, the examples on this page are pretty fiddly. The most important bit is actually getting the sketch right.
If you don't, you're in big trouble. Then you've just got to carefully use the sketch to work out the other solutions.
It's tricky, but you'll feel better about yourself when you've got it mastered. Ah you will, you will, you will ...

Solving Trig Equations in a Given Interval

Now for something really exciting — trig identities. Mmm, well, maybe exciting was the wrong word.
But they can be dead useful, so here goes...

For equations with **tan x** in, it often helps to use this...

$$\tan x \equiv \frac{\sin x}{\cos x}$$

This is a handy thing to know — and one the examiners love testing. Basically, if you've got a trig equation with a tan in it, together with a sin or a cos — chances are you'll be better off if you rewrite the tan using this formula.

EXAMPLE: Solve: $3\sin x - \tan x = 0$, for $0 \leq x \leq 2\pi$.

It's got sin and tan in it — so writing tan x as $\dfrac{\sin x}{\cos x}$ is probably a good move:

$$3\sin x - \tan x = 0$$
$$\Rightarrow 3\sin x - \frac{\sin x}{\cos x} = 0$$

Get rid of the cos x on the bottom by multiplying the whole equation by cos x.

$$\Rightarrow 3\sin x \cos x - \sin x = 0$$

Now — there's a common factor of sin x. Take that outside a bracket.

$$\Rightarrow \sin x (3\cos x - 1) = 0$$

And now you're almost there. You've got two things multiplying together to make zero. That means either one or both of them is equal to zero themselves.

$$\Rightarrow \sin x = 0 \quad \text{or} \quad 3\cos x - 1 = 0$$

CAST gives any solutions in the interval $0 \leq x \leq 2\pi$.

$\sin x = 0$

The first solution is... $\sin 0 = 0$

Now find the other points where sin x is zero in the interval $0 \leq x \leq 2\pi$.
(Remember the sin graph is zero every π radians.)

$$\Rightarrow x = 0, \pi, 2\pi \text{ radians}$$

So altogether you've got <u>five</u> possible solutions:

$$\Rightarrow x = 0, \pi, 2\pi, 1.231, 5.052 \text{ radians}$$

$3\cos x - 1 = 0$

Rearrange... $\cos x = \frac{1}{3}$

So the first solution is...

$$\cos^{-1}\frac{1}{3} = 1.231$$

CAST (or the graph of cos x) gives another solution in the 4th quadrant...

And the two solutions from this part are:

$$\Rightarrow x = 1.231, 5.052 \text{ radians}$$

Trigonometry is the root of all evil...

What a page — you don't have fun like that every day, do you? No, trig equations are where it's at. This is a really useful trick, though — and can turn a nightmare of an equation into a bit of a pussy-cat. Rewriting stuff using different formulas is always worth trying if it feels like you're getting stuck — even if you're not sure why when you're doing it. You might have a flash of inspiration when you see the new version.

Solving Trig Equations in a Given Interval

Another trig identity — and it's a good 'un — examiners love it. And it's not difficult either.

And if you have a *sin² x* or a *cos² x*, think of this straight away...

$$\sin^2 x + \cos^2 x \equiv 1 \implies \begin{array}{l} \sin^2 x \equiv 1 - \cos^2 x \\ \cos^2 x \equiv 1 - \sin^2 x \end{array}$$

Use this identity to get rid of a sin² or a cos² that's making things awkward...

EXAMPLE: Solve: $2\sin^2 x + 5\cos x = 4$, for $0° \leq x \leq 360°$.

You can't do much while the equation's got both sin's and cos's in it. So replace the sin²x bit with 1 – cos²x.

$$2(1 - \cos^2 x) + 5\cos x = 4$$

Multiply out the bracket and rearrange it so that you've got zero on one side — and you get a quadratic in cos x:

Now the only trig function is cos.

$$\Rightarrow 2 - 2\cos^2 x + 5\cos x = 4$$
$$\Rightarrow 2\cos^2 x - 5\cos x + 2 = 0$$

If you replaced cos x with y, this would be $2y^2 - 5y + 2 = 0$.

This is a quadratic in cos x. It's easier to factorise this if you make the substitution y = cos x.

$$2y^2 - 5y + 2 = 0$$
$$\Rightarrow (2y - 1)(y - 2) = 0$$
$$\Rightarrow (2\cos x - 1)(\cos x - 2) = 0$$

$2y^2 - 5y + 2 = (2y \ ?)(y \ ?)$
$= (2y - 1)(y - 2)$

Now one of the brackets must be 0. So you get 2 equations as usual:

You've already done this example on page 12.

$$(2\cos x - 1) = 0 \quad or \quad (\cos x - 2) = 0$$

This is a bit weird. cos x is always between –1 and 1. So you don't get any solutions from this bracket.

$$\cos x = \tfrac{1}{2} \Rightarrow x = 60° \quad or \quad x = 300° \quad and \quad \boxed{\cos x = 2}$$ This is impossible — so you get nothing from this bracket.

So at the end of all that, the only solutions you get are x = 60° and x = 300°. How boring.

Use the **Trig Identities** to prove something is the **Same** as something else

Another use for these trig identities is proving that two things are the same.

EXAMPLE: Show that $\dfrac{\cos^2 \theta}{1 + \sin \theta} \equiv 1 - \sin \theta$

The identity sign ≡ means that this is true for all θ, rather than just certain values.

Prove things like this by playing about with one side of the equation until you get the other side.

<u>Left-hand side</u>: $\dfrac{\cos^2 \theta}{1 + \sin \theta}$

The only thing I can think of doing here is replacing cos² θ with 1 – sin² θ. (Which is good because it works.)

$$\equiv \dfrac{1 - \sin^2 \theta}{1 + \sin \theta}$$

The next trick is the hardest to spot. Look at the top — does that remind you of anything?

The top line is a difference of two squares:

$$\equiv \dfrac{(1 + \sin \theta)(1 - \sin \theta)}{1 + \sin \theta}$$

$1 - a^2 = (1 + a)(1 - a)$
$\Rightarrow 1 - \sin^2 \theta = (1 + \sin \theta)(1 - \sin \theta)$

$$\equiv 1 - \sin \theta, \text{ the right-hand side.}$$

Trig identities — the path to a brighter future...

That was a pretty miserable section. But it's over. These trig identities aren't exactly a barrel of laughs, but they are a definite source of marks — you can bet your last penny they'll be in the exam. That substitution trick to get rid of a sin² or a cos² and end up with a quadratic in sin x or cos x is a real examiners' favourite. Those identities can be a bit daunting, but it's always worth having a few tricks in the back of your mind — always look for things that factorise, or fractions that can be cancelled down, or ways to use those trig identities. Ah, it's all good clean fun.

Section Two Revision Questions

Welcome to <u>Who wants to be a Mathematician</u> — the page that gives you the chance to win a million marks in your exam by answering questions on a topic of your choice. You have chosen Section Two — Trigonometry...

For £100: Draw a triangle $\triangle XYZ$ with sides of length x, y and z. Write down the Sine and Cosine Rules for this triangle. Write down an expression for its area.

For £200: Write down the exact values of cos 30°, sin 30°, tan 30°, cos 45°, sin 45°, tan 45°, cos 60°, sin 60°, tan 60°. (You'll probably want to draw a couple of triangles to help you.)

For £300: What is tan x in terms of cos x and sin x? What is $\cos^2 x$ in terms of $\sin^2 x$?

For £500: Solve a) $\triangle ABC$ in which A = 30°, C = 25°, b = 6m and find its area.
b) $\triangle PQR$ in which p = 3 km, q = 23 km, R = 10°.   (answers to 2 d.p.)

For £1000: My pet triangle Freda has sides of length 10, 20, 25. Find her angles (to 1 d.p.) and sketch her.

Well done — you now have a guaranteed win of £1000.

For £2000: Find the 2 possible triangles $\triangle ABC$ which satisfy b = 5, a = 3, A = 35°. (This is tricky: Sketch it first, and see if you can work out how to make 2 different triangles satisfying the data given.)

For £4000: Sketch the graphs for sin x, cos x and tan x.
Make sure you label all the max/min/zero/undefined points.

For £8000: Sketch the following graphs:

     a) $y = \frac{1}{2} \cos 2x$ (for $0° \le x \le 360°$)   b) $y = 2 \sin (x+30°)$ (for $0° \le x \le 360°$)

     c) $y = \tan 3x$ (for $0° \le x \le 180°$)

For £16 000: i) Solve each of these equations for $0° \le \theta \le 360°$:

     a) $\sin\theta = -\frac{\sqrt{3}}{2}$ b) $\tan\theta = -1$ c) $\cos\theta = -\frac{1}{\sqrt{2}}$

     ii) Solve each of these equations for $-180° \le \theta \le 180°$ (giving your answer to 1 d.p.):

     a) $\cos 4\theta = -\frac{2}{3}$ b) $\sin(\theta + 35°) = 0.3$ c) $\tan(\frac{1}{2}\theta) = 500$

For £32 000: Give the radius and the coordinates of the centre of the circles with the following equations:

     a) $x^2 + y^2 = 9$ b) $(x-2)^2 + (y+4)^2 = 4$ c) $x(x+6) = y(8-y)$

Congratulations! You will now be taking home a cheque for at least £32000!

For £64 000: Find all the solutions to $6\sin^2 x = \cos x + 5$ in the range $0° \le x \le 400°$ (answers to 1 d.p.).

For £125 000: Solve $3\tan x + 2\cos x = 0$ for $-90° \le x \le 90°$

For £250 000: Simplify: $(\sin y + \cos y)^2 + (\cos y - \sin y)^2$

For £500 000: Show that $\dfrac{\sin^4 x + \sin^2 x\cos^2 x}{\cos^2 x - 1} \equiv -1$

Here comes the big one. Are you ready?

For £1 million: Which of the following identities is correct?
     A $\sin^2 x + \cos^2 x \equiv \tan^2 x$ B $\sin^2 x - \cos^2 x \equiv 1$

     C $\sin^2 ☺ + \cos^2 ♟ \equiv ☙$ D $\sin^2 x + \cos^2 x \equiv 1$

Are you sure? Is that your final answer? Perhaps you'd like to phone a teacher.

Well I can tell you, if you <u>had</u> said C........

.... you would have just <u>lost</u> £468,000...

Logs

Don't be put off by your parents or grandparents telling you that logs are hard. <u>Logarithm</u> is just a fancy word for <u>power</u>, and once you know how to use them you can solve all sorts of equations.

You need to be able to **Switch** between **Different Notations**

> $\log_a b = c$ means the same as $a^c = b$
>
> That means that $\log_a a = 1$ and $\log_a 1 = 0$

The little number 'a' after 'log' is called the <u>base</u>. Logs can be to any base, but <u>base 10</u> is the most common. The <u>button</u> marked '<u>log</u>' on your calculator uses base 10.

Example: Index notation: $10^2 = 100$ log notation: $\log_{10} 100 = 2$

The <u>base</u> goes here but it's usually left out if it's 10.

So the <u>logarithm</u> of 100 to the <u>base 10</u> is 2, because 10 raised to the <u>power</u> of 2 is 100.

Examples:

Write down the values of the following:

a) $\log_2 8$ b) $\log_9 3$ c) $\log_5 5$

a) 8 is 2 raised to the power of 3
so $2^3 = 8$ and $\log_2 8 = 3$

b) 3 is the square root of 9, or $9^{1/2} = 3$
so $\log_9 3 = \frac{1}{2}$

c) anything to the power of 1 is itself
so $\log_5 5 = 1$

Write the following using log notation:

a) $5^3 = 125$ b) $3^0 = 1$

You just need to make sure you get things in the right place.

a) 3 is the power or <u>logarithm</u> that 5 (the <u>base</u>) is raised to to get 125
so $\log_5 125 = 3$

b) you'll need to remember this one:
$\log_3 1 = 0$

The **Laws of Logarithms** are **Unbelievably Useful**

Whenever you have to deal with <u>logs</u>, you'll end up using the <u>laws</u> below. That means it's no bad idea to <u>learn them</u> off by heart right now.

Laws of Logarithms

$$\log_a x + \log_a y = \log_a (xy)$$
$$\log_a x - \log_a y = \log_a \left(\frac{x}{y}\right)$$
$$\log_a x^k = k \log_a x$$

So $\log_a \frac{1}{x} = -\log_a x$

You've got to be able to change the <u>base</u> of a log too.

So $\log_7 4 = \dfrac{\log_{10} 4}{\log_{10} 7} = 0.7124$

To check: $7^{0.7124} = 4$

Change of Base

$$\log_a x = \frac{\log_b x}{\log_b a}$$

Use the **Laws** to **Manipulate Logs**

Example: Write each expression in the form $\log_a n$, where n is a number.
a) $\log_a 5 + \log_a 4$ b) $2 \log_a 6 - \log_a 9$

a) $\log_a x + \log_a y = \log_a (xy)$

You just have to <u>multiply</u> the numbers together:

$\log_a 5 + \log_a 4 = \log_a (5 \times 4)$
$= \log_a 20$

b) $\log_a x^k = k \log_a x$

$2 \log_a 6 = \log_a 6^2 = \log_a 36$
$\log_a 36 - \log_a 9 = \log_a (36 \div 9)$
$= \log_a 4$

It's sometimes hard to see the wood for the trees — especially with logs...

Tricky, tricky, tricky... I think of $\log_a b$ as 'the <u>power</u> I have to raise a to if I want to end up with b' — that's all it is. And the log laws make a bit more sense if you think of 'log' as meaning 'power'. For example, you know that $2^a \times 2^b = 2^{a+b}$ — this just says that if you multiply two numbers, you add the powers. Well, the first law of logs is saying the same thing. Any road up, even if you don't really understand why they work, make sure you know the log laws like you know your own navel.

Exponentials and Logs

Okay, you've done the theory of logs. So now it's a bit of stuff about exponentials (the opposite of logs, kind of), and then it'll be time to get your calculator out for a bit of button pressing...

Graphs of a^x Never Reach Zero

All the graphs of $y = a^x$ (exponential graphs) where $a > 1$ have the <u>same basic shape</u>. The graphs for $a = 2$, $a = 3$ and $a = 4$ are shown on the right.

- All the a's are greater than 1 — so <u>y increases as x increases</u>.
- The <u>bigger</u> a is, the <u>quicker</u> the graphs increase. The rate at which they increase gets bigger too.
- As x <u>decreases</u>, y <u>decreases</u> at a <u>smaller and smaller rate</u> — y will approach zero, but never actually get there.

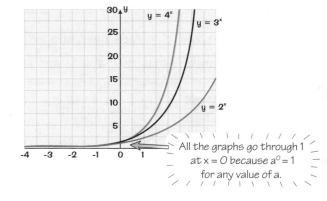

All the graphs go through 1 at x = O because $a^0 = 1$ for any value of a.

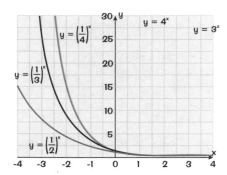

The graphs on the left are for $y = a^x$ where $a < 1$ (they're for $a = \frac{1}{2}$, $\frac{1}{3}$ and $\frac{1}{4}$).

- All the a's are less than 1 — meaning <u>y decreases as x increases</u>.
- As x <u>increases</u>, y <u>decreases</u> at a <u>smaller and smaller rate</u> — again, y will approach zero, but never actually get there.

You can use **exponentials** and **logs** to **solve equations**

Example:

1) Solve $2^{4x} = 3$ to 3 decimal places.

You want x on its own, so take logs of both sides (by writing 'log' in front of both sides):

$\log 2^{4x} = \log 3$

Now use one of the laws of logs: $\log x^k = k \log x$:

$4x \log 2 = \log 3$

You can now divide both sides by '4 log 2' to get x on its own:

$x = \dfrac{\log 3}{4 \log 2}$

But $\dfrac{\log 3}{4 \log 2}$ is just a number you can find using a calculator:

$x = 0.396$ (to 3 d.p.)

I don't know about you, but I enjoyed that more than the biggest, fastest rollercoaster. You want another? OK then...

Example:

2) Solve $7 \log_{10} x = 5$ to 3 decimal places.

You want x on its own, so begin by dividing both sides by 7:

$\log_{10} x = \frac{5}{7}$

You now need to take exponentials of both sides by doing '10 to the power of both sides' (since the log is to base 10):

$10^{\log_{10} x} = 10^{\frac{5}{7}}$

logs and exponentials are inverse functions, so they cancel out:

$x = 10^{\frac{5}{7}}$

Again, $10^{\frac{5}{7}}$ is just a number you can find using a calculator:

$x = 5.179$

Exponentials and Logs

Use the **Calculator Log Button** Whenever You Can

Example: Use logarithms to solve the following for x, giving the answers to 4 s.f.
a) $10^x = 170$ b) $10^{3x} = 4000$ c) $7^x = 55$ d) $\log_{10}x = 2.6$ e) $\log_2 x = 5$

You've got the magic buttons on your calculator, but you'd better <u>follow the instructions</u> and show that you know how to use the <u>log rules</u> covered earlier.

a) $10^x = 170$ — there's an 'unknown' in the power, so <u>take logs of both sides</u>.
(In theory, it doesn't matter what <u>base</u> you use, but your calculator has a '$\log_{10}$' button, so base 10 is usually a good idea. But whatever base you use, <u>use the same one for both sides</u>.)
So taking logs to base 10 of both sides of the above equation gives:

$$\log 10^x = \log 170$$
$$\text{i.e. } x\log 10 = \log 170$$

Since $\log_{10}10 = 1$ ⟹ i.e. $x = \log 170 = 2.230$ (to 4 sig. fig)

b) $10^{3x} = 4000$. Same again — take logs (to base 10) of both sides to get: $3x = \log_{10} 4000 = 3.602$, so x = 1.201

c) $7^x = 55$. Once again, take logs of both sides, and use the log rules: $x \log_{10}7 = \log_{10}55$, so $x = \dfrac{\log_{10} 55}{\log_{10} 7} = 2.059$

d) $\log_{10}x = 2.6$ — to get rid of a log, you 'take exponentials', meaning you do '10 (or the base) to the power of each side'.
Think of 'taking logs' and 'taking exponentials' as opposite processes — one cancels the other out: $10^{\log_{10} x} = 10^{2.6}$

Or you can simply use the formula on p17 — whichever you prefer ⟹ i.e. $x = 398.1$

e) $\log_2 x = 5$. Here, the base is 2, so 'taking exponentials' means doing '2 to the power of both sides': $2^{\log_2 x} = 2^5$
i.e. $x = 2^5 = 32$

Exponential Growth and **Decay** Applies to **Real-life** Problems

Logs can even be used to solve real-life problems.

Example: The radioactivity of a substance decays by 20 per cent over a year. The initial level of radioactivity is 400. Find the time taken for the radioactivity to fall to 200 (the half-life).

$R = 400 \times 0.8^T$ where R is the <u>level of radioactivity</u> at time T years.
We need R = 200 so solve $200 = 400 \times 0.8^T$

The 0.8 comes from 1 − 20% decay.

$0.8^T = \dfrac{200}{400} = 0.5.$

$T \log 0.8 = \log 0.5.$

$T = \dfrac{\log 0.5}{\log 0.8} = 3.106$ years

If in doubt, take the log of something — that usually works...

The thing about exponential growth is that it's really useful, as it happens in real life all over the place. Money in a bank account earns interest at <u>a certain percentage per year</u>, and so the balance rises <u>exponentially</u> (if you don't spend or save anything). Likewise, if you got 20% cleverer for every week you studied, you'd eventually end up infinitely clever.

Section Three Revision Questions

Logs* and exponentials are surprisingly useful things. As well as being in your exam they pop up all over the place in real life — savings, radioactive decay, growth of bacteria — all log*arithmic.

And now for something (marginally) different:

1) Write down the values of the following:
   a) $\log_3 27$
   b) $\log_3 (1 \div 27)$
   c) $\log_3 18 - \log_3 2$

2) Simplify the following:
   a) $\log 3 + 2 \log 5$
   b) $\frac{1}{2} \log 36 - \log 3$

3) Simplify $\log_b (\chi^2 - 1) - \log_b (\chi - 1)$

4) a) Copy and complete the table for the function $y = 4^x$:

x	−3	−2	−1	0	1	2	3
y							

   b) Using suitable scales, plot a graph of $y = 4^x$ for $-3 < x < 3$.
   c) Use the graph to solve the equation $4^x = 20$.

5) Solve these little jokers:
   a) $10^x = 240$
   b) $\log_{10} x = 2.6$
   c) $10^{2x+1} = 1500$
   d) $4^{(x-1)} = 200$

6) Find the smallest integer P such that $1.5^P > 1\,000\,000$

*   He's a lumberjack and he's okay.
    He sleeps all night and he works all day.

Geometric Progressions

You have a geometric progression when each term in a sequence is found by multiplying the previous term by a (constant) number. Let me explain...

Geometric Progressions Multiply by a Constant each time

Geometric progressions work like this: the next term in the sequence is obtained by multiplying the previous one by a constant value. Couldn't be easier.

$$u_1 = a \qquad\qquad = a$$
$$u_2 = a \times r \qquad\quad = ar$$
$$u_3 = a \times r \times r \qquad = ar^2$$
$$u_4 = a \times r \times r \times r = ar^3$$

The first term (u_1) is called 'a'.

The number you multiply by each time is called 'the common ratio', symbolised by 'r'.

Here's the formula describing any term in the geometric progression:

$$u_n = ar^{n-1}$$

Example: There is a chessboard with a 1p piece on the first square, 2p on the second square, 4p on the third, 8p on the fourth and so on until the board is full. Calculate how much money is on the board.

This is a geometric progression, where you get the next term in the sequence by multiplying the previous one by 2.

So a = 1 (because you start with 1p on the first square) and r = 2.

So $u_1 = 1$, $u_2 = 2$, $u_3 = 4$, $u_4 = 8$, ...

A Sequence becomes a Series when you Add the Terms to find the Total

S_n stands for the sum of the first n terms of the geometric progression.
In the example above, you're told to work out S_{64} (because there are 64 squares on a chessboard).

To work out the formula for the sum of a G.P. you use two series and subtract.

For a G.P.:	$S_n = a + ar + ar^2 + ar^3 + \ldots + ar^{n-1}$
Multiplying by r gives:	$rS_n = ar + ar^2 + ar^3 + \ldots + ar^{n-2} + ar^{n-1} + ar^n$
Subtracting gives:	$S_n - rS_n = a - ar^n$
Factorising:	$(1-r)S_n = a(1-r^n) \implies \boxed{S_n = \dfrac{a(1-r^n)}{1-r}}$

If the series were subtracted the other way around you'd get
$$S_n = \frac{a(r^n - 1)}{r - 1}.$$
Both versions are correct.

So, back to the chessboard example:

a = 1, r = 2, n = 64

$$S_{64} = \frac{1(1-2^{64})}{1-2}$$
$$S_{64} = 1.84 \times 10^{19} \text{ pence or } £1.84 \times 10^{17}$$

The whole is more than the sum of the parts — hmm, not in maths, it ain't...

You really need to understand the difference between arithmetic and geometric progressions — it's not hard, but it needs to be fixed firmly in your head. There are only a few formulas for sequences and series (the nth term of a sequence, the sum of the first n terms of a series), but you need to learn them, since they won't be in the formula book they give you.

Geometric Progressions

Geometric progressions can either Grow or Shrink

In the chessboard example, each term was <u>bigger</u> than the previous one, 1, 2, 4, 8, 16, ...
You can create a series where each term is <u>less</u> than the previous one by using a <u>small value of r</u>.

Example: If $a = 20$ and $r = \frac{1}{5}$, write down the first five terms of the sequence and the 20th term.

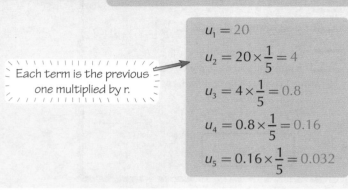

$u_1 = 20$

$u_2 = 20 \times \frac{1}{5} = 4$

Each term is the previous one multiplied by r.

$u_3 = 4 \times \frac{1}{5} = 0.8$

$u_4 = 0.8 \times \frac{1}{5} = 0.16$

$u_5 = 0.16 \times \frac{1}{5} = 0.032$

$u_{20} = ar^{19}$

$= 20 \times \left(\frac{1}{5}\right)^{19}$

$= 1.048576 \times 10^{-12}$

The sequence is <u>tending towards zero</u>, but won't ever get there.

In general, for each term to be <u>smaller</u> than the one before, you need $|r| < 1$.
A sequence with $|r| < 1$ is called <u>convergent</u>, since the terms converge to a limit.
Any other sequence (like the chessboard example on page 21) is called <u>divergent</u>.

$|r|$ means the modulus (or size) of r, <u>ignoring the sign</u> of the number. So $|r| < 1$ means that $-1 < r < 1$.

A Convergent Series has a Sum to Infinity

In other words, if you just <u>kept</u> adding terms to a <u>convergent series</u>, you'd get <u>closer and closer</u> to a certain number, but you'd never actually reach it.

If $|r| < 1$ and n is very, very <u>big</u>, then r^n will be very, very <u>small</u> — or to put it technically, $r^n \to 0$.
(Try working out $(\frac{1}{2})^{100}$ on your calculator if you don't believe me.)
This means $(1 - r^n)$ is really, really close to 1.

So, as $n \to \infty$, $S_n \to \dfrac{a}{1-r}$.

It's easier to remember as $S_\infty = \dfrac{a}{1-r}$

S_∞ just means 'sum to infinity'.

Example: If $a = 2$ and $r = \frac{1}{2}$, find the sum to infinity of the geometric series.

$u_1 = 2$ $\longrightarrow$ $S_1 = 2$

$u_2 = 2 \times \frac{1}{2} = 1$ $\longrightarrow$ $S_2 = 2 + 1 = 3$

These values are getting <u>smaller</u> each time.

$u_3 = 1 \times \frac{1}{2} = \frac{1}{2}$ $\longrightarrow$ $S_3 = 2 + 1 + \frac{1}{2} = 3\frac{1}{2}$

$u_4 = \frac{1}{2} \times \frac{1}{2} = \frac{1}{4}$ $\longrightarrow$ $S_4 = 2 + 1 + \frac{1}{2} + \frac{1}{4} = 3\frac{3}{4}$

$u_5 = \frac{1}{4} \times \frac{1}{2} = \frac{1}{8}$ $\longrightarrow$ $S_5 = 2 + 1 + \frac{1}{2} + \frac{1}{4} + \frac{1}{8} = 3\frac{7}{8}$

$u_6 = \frac{1}{8} \times \frac{1}{2} = \frac{1}{16}$ $\longrightarrow$ $S_6 = 2 + 1 + \frac{1}{2} + \frac{1}{4} + \frac{1}{8} + \frac{1}{16} = 3\frac{15}{16}$

These values are getting closer (<u>converging</u>) to 4.
So, the sum to infinity is 4.

You can show this <u>graphically</u>:
The line on the graph is getting <u>closer and closer</u> to 4, but it'll never actually get there.

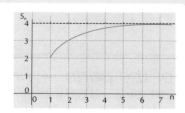

Of course, you could have saved yourself a lot of bother by using the <u>sum to infinity formula</u>:

$S_\infty = \dfrac{a}{1-r} = \dfrac{2}{1-\frac{1}{2}} = 4$

Geometric Progressions

A *Divergent* series *Doesn't* have a sum to infinity

Example: If $a = 2$ and $r = 2$, find the sum to infinity of the series.

$$u_1 = 2 \implies S_1 = 2$$
$$u_2 = 2 \times 2 = 4 \implies S_2 = 2 + 4 = 6$$
$$u_3 = 4 \times 2 = 8 \implies S_3 = 2 + 4 + 8 = 14$$
$$u_4 = 8 \times 2 = 16 \implies S_4 = 2 + 4 + 8 + 16 = 30$$
$$u_5 = 16 \times 2 = 32 \implies S_5 = 2 + 4 + 8 + 16 + 32 = 62$$

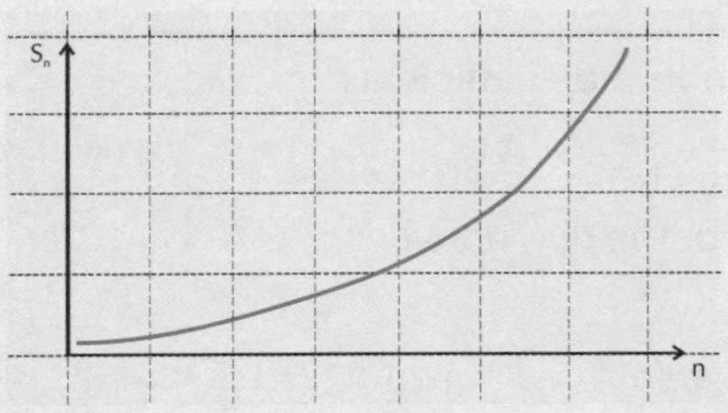

This is an exponential graph — see section 3.

As $n \to \infty$, $S_n \to \infty$ in a big way. So big, in fact, that eventually you can't work it out — so don't bother.

There is no sum to infinity for a divergent series.

Example: When a baby is born, £3000 is invested in an account with a fixed interest rate of 4% per year.
a) What will the account be worth at the start of the seventh year?
b) Will the account have doubled in value by the time the child reaches its 21st birthday?

a) $u_1 = a = 3000$

$u_2 = 3000 + (4\% \text{ of } 3000)$ ← This is the interest.
$= 3000 + (0.04 \times 3000)$
$= 3000 (1 + 0.04)$
$= 3000 \times 1.04$ ← So, $r = 1.04$

$u_3 = u_2 \times 1.04$
$= (3000 \times 1.04) \times 1.04$
$= 3000 \times (1.04)^2$

$u_4 = 3000 \times (1.04)^3$

I've missed out some steps here — check that you understand what's happened.

$u_7 = 3000 \times (1.04)^6$
$= £3795.96$ (to the nearest penny)

b) You need to know when $u_n > 6000$ ← double the original value.
From part a) you can tell that $u_n = 3000 \times (1.04)^{n-1}$
So $3000 \times (1.04)^{n-1} > 6000$
$(1.04)^{n-1} > 2$
To complete this you need to use logs (see section 3):
$\log(1.04)^{n-1} > \log 2$
$(n-1)\log(1.04) > \log 2$
$n - 1 > \dfrac{\log 2}{\log 1.04}$
$n - 1 > 17.67$
$n > 18.67$ (to 2 d.p.)

So u_{19} (the amount at the start of the 19th year) will be more than double the original amount — plenty of time to buy a Porsche for the 21st birthday.

So tell me — if my savings earn 4% per year, when will I be rich...

Now here's a funny thing — you can have a convergent geometric series if the common ratio is small enough.
I find this odd — that I can keep adding things to a sum forever, but the sum never gets really really big.

Binomial Expansions

If you're feeling a bit stressed, just take a couple of minutes to relax before trying to get your head round this page — it's a bit of a stinker in places. Have a cup of tea and think about something else for a couple of minutes. Ready...

Writing *Binomial Expansions* is all about *Spotting Patterns*

Doing binomial expansions just involves <u>multiplying out</u> the brackets. It would get nasty when you raise the brackets to <u>higher powers</u> — but once again I've got a <u>cunning plan</u>...

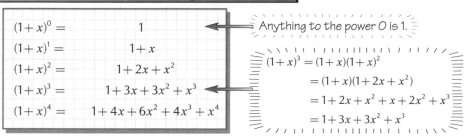

$$(1+x)^0 = 1$$
$$(1+x)^1 = 1+x$$
$$(1+x)^2 = 1+2x+x^2$$
$$(1+x)^3 = 1+3x+3x^2+x^3$$
$$(1+x)^4 = 1+4x+6x^2+4x^3+x^4$$

Anything to the power 0 is 1.

$$(1+x)^3 = (1+x)(1+x)^2$$
$$= (1+x)(1+2x+x^2)$$
$$= 1+2x+x^2+x+2x^2+x^3$$
$$= 1+3x+3x^2+x^3$$

A Frenchman named Pascal spotted the pattern in the coefficients and wrote them down in a <u>triangle</u>.
So it was called 'Pascal's Triangle' (imaginative, eh?).
The pattern's easy — each number is the <u>sum</u> of the two above it.

So, the next line will be: **1 5 10 10 5 1**
giving **(1 + x)⁵ = 1 + 5x + 10x² + 10x³ + 5x⁴ + x⁵.**

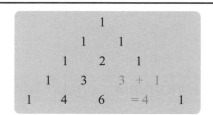

You *Don't* need to write out Pascal's Triangle for *Higher Powers*

There's a formula for the numbers in the triangle. The formula looks <u>horrible</u> (one of the worst in AS maths) so don't try to learn it letter by letter — look for the <u>patterns</u> in it instead. Here's an example:

Example: Expand $(1 + x)^{20}$, giving the first four terms only.

So you can use this formula for any power, the power is called n. In this example n = 20.

$$(1+x)^n = 1 + \frac{n}{1}x + \frac{n(n-1)}{1 \times 2}x^2 + \boxed{\frac{n(n-1)(n-2)}{1 \times 2 \times 3}x^3} + \ldots\ldots + x^n$$

Here's a closer look at the term in the black box:

There are <u>three things</u> multiplied together on the top row. If n=20, this would be 20×19×18. $\longrightarrow$ $\dfrac{n(n-1)(n-2)}{1 \times 2 \times 3}x^3$ $\longleftarrow$ <u>Start here</u>. The power of x is 3 and everything else here is based on 3.

There are <u>three integers</u> here multiplied together.
1×2×3 is written as 3! and called 3 <u>factorial</u>.

This means, if n = 20 and you were asked for '<u>the term in x⁷</u>' you should write $\dfrac{20 \times 19 \times 18 \times 17 \times 16 \times 15 \times 14}{1 \times 2 \times 3 \times 4 \times 5 \times 6 \times 7}x^7$.

This can be <u>simplified</u> to $\dfrac{20!}{7!13!}x^7$ $\longleftarrow$ $20 \times 19 \times 18 \times 17 \times 16 \times 15 \times 14 = \dfrac{20!}{13!}$ because it's the numbers from 20 to 1 multiplied together, divided by the numbers from 13 to 1 multiplied together.

Believe it or not, there's an even <u>shorter</u> form: $\dfrac{20!}{7!13!}$ is written as $^{20}C_7$ or $\dbinom{20}{7}$ $^nC_r = \dbinom{n}{r} = \dfrac{n!}{r!(n-r)!}$

So, to finish the example, $(1+x)^{20} = 1 + \dfrac{20}{1}x + \dfrac{20 \times 19}{1 \times 2}x^2 + \dfrac{20 \times 19 \times 18}{1 \times 2 \times 3}x^3 + \ldots$ $= 1 + 20x + 190x^2 + 1140x^3 + \ldots$

Binomial Expansions

It's slightly more complicated when the Coefficient of x isn't 1

Example: What is the term in x^5 in the expansion of $(1 - 3x)^{12}$?

The term in x^5 will be as follows:

$$\frac{12 \times 11 \times 10 \times 9 \times 8}{1 \times 2 \times 3 \times 4 \times 5}(-3x)^5$$

$$= \frac{12!}{5!7!}(-3)^5 x^5 \quad = -\frac{12!}{5!7!} \times 3^5 x^5 = -192456x^5$$

Watch out — the –3 is included here with the x.

Tip — the digits on the bottom of the fraction should always add up to the number on the top.

Note that $(-3)^{even}$ will always be positive and $(-3)^{odd}$ will always be negative.

Some Binomials contain More Complicated Expressions

The binomials so far have all had a 1 in the brackets — things get tricky when there's a number other than 1. Don't panic, though. The method is the same as before once you've done a bit of factorising.

Example: What is the coefficient of x^4 in the expansion of $(2 + 5x)^7$?

Factorising $(2 + 5x)$ gives $2(1 + \frac{5}{2}x)$

So, $(2 + 5x)^7$ gives $2^7(1 + \frac{5}{2}x)^7$

It's really easy to forget the first bit (here it's 2^7) — you've been warned...

Here's the one you want.

$$(2 + 5x)^7 = 2^7(1 + \frac{5}{2}x)^7$$

$$= 2^7[1 + 7\left(\frac{5}{2}x\right) + \frac{7 \times 6}{1 \times 2}\left(\frac{5}{2}x\right)^2 + \frac{7 \times 6 \times 5}{1 \times 2 \times 3}\left(\frac{5}{2}x\right)^3 + \frac{7 \times 6 \times 5 \times 4}{1 \times 2 \times 3 \times 4}\left(\frac{5}{2}x\right)^4 + ...]$$

The coefficient of x^4 will be $2^7 \times \frac{7!}{4!3!}(\frac{5}{2})^4 = 175000$

Don't forget the 2^7.

So, there's no need to work out all of the terms.
In fact, you could have gone directly to the term in x^4 by using the method on page 24.

Note: The question asked for the coefficient of x^4 in the expansion, so don't include any x's in your answer. If you'd been asked for the term in x^4 in the expansion, then you should have included the x^4 in your answer.
Always read the question very carefully.

Pascal was fine at maths but rubbish at music — he only played the triangle...

You can use your calculator to work out these tricky fractions — you use the ^nC_r button (though it could be called something else on your calculator). So to work out $^{20}C_7$, press '20', then press the ^nC_r button, then press '7', and then finish with '='. Now work out $^{15}C_7$ and $^{15}C_8$ — you should get the same answers, since they're both $\frac{15!}{7!8!}$.

Section Four Revision Questions

What's that I hear you cry? You want revision questions — and lots of them. Well it just so happens I've got a few here. Various questions on geometric series and binomial expansions.

With sequences and series, get a clear idea in your head before you start, or you could get to the end of the question and realise you've been heading in completely the wrong direction. That would be bad.

1) For the geometric progression 2, –6, 18, ..., find:
   a) the 10th term,
   b) the sum of the first 10 terms.

2) Find the sum of the first 12 terms of the following geometric series:
   a) $2 + 8 + 32 + ...$
   b) $30 + 15 + 7.5 + ...$

3) Find the common ratio for the following geometric series.
   State which ones are convergent and which are divergent.
   a) $1 + 2 + 4 + ...$
   b) $54 + 27 + 9 + ...$
   c) $1 + \dfrac{1}{3} + \dfrac{1}{9} + ...$
   d) $4 + 1 + \dfrac{1}{4} + ...$

4) For the geometric progression 24, 12, 6, ..., find:
   a) the common ratio,
   b) the seventh term,
   c) the sum of the first 10 terms,
   d) the sum to infinity.

5) A geometric progression begins 2, 6, ...
   Which term of the geometric progression equals 1458?

6) Write down the sixth row of Pascal's Triangle. (Hint: it starts with a '1'.)

7) Give the first four terms in the expansion of $(1 + x)^{12}$.

8) What is the term in x^4 in the expansion of $(1 - 2x)^{16}$?

9) Find the coefficient of x^2 in the expansion of $(2 + 3x)^5$.

Stationary Points

Differentiation is how you find gradients of curves. So you can use differentiation to find a <u>stationary point</u> (where a graph 'levels off') — that means finding where the <u>gradient</u> becomes <u>zero</u>.

Stationary Points are when the gradient is *Zero*

Example: Find the stationary points on the curve $y = 2x^3 - 3x^2 - 12x + 5$, and work out the nature of each one.

A <u>stationary point</u> can be...

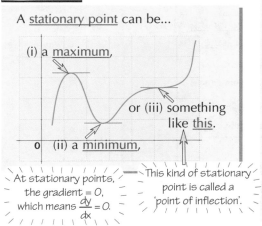

(i) a <u>maximum</u>,

or (iii) something like <u>this</u>.

(ii) a <u>minimum</u>,

At stationary points, the gradient = 0, which means $\frac{dy}{dx} = 0$.

This kind of stationary point is called a 'point of inflection'.

You need to find where $\frac{dy}{dx} = f'(x) = 0$. So first, <u>differentiate</u> the function.

$$y = 2x^3 - 3x^2 - 12x + 5 \Rightarrow \frac{dy}{dx} = 6x^2 - 6x - 12$$

This is the expression for the gradient.

And then set this derivative equal to <u>zero</u>.

$f'(x)$ (pronounced "f-dash of x") is just another way to write a derivative.

$$6x^2 - 6x - 12 = 0$$
$$\Rightarrow x^2 - x - 2 = 0$$
$$\Rightarrow (x-2)(x+1) = 0$$
$$\Rightarrow x = 2 \text{ or } x = -1$$

So the graph has <u>two</u> stationary points, at $x = 2$ and $x = -1$.

The stationary points are actually at $(2, -15)$ and $(-1, 12)$.

Substitute the x values into the function to find the y-coordinates.

Decide if it's a *Maximum* or a *Minimum* by differentiating *Again*

Once you've found where the stationary points are, you have to decide whether each of them is a <u>maximum</u> or <u>minimum</u> — this is all a question means when it says, '...determine the <u>nature</u> of the turning points'.

A turning point is a another name for a maximum or a minimum.

To decide whether a stationary point is a <u>maximum</u> or a <u>minimum</u> — just differentiate again to find $\frac{d^2y}{dx^2}$ (or $f''(x)$.)

If $\frac{d^2y}{dx^2} < 0$, it's a <u>maximum</u>.

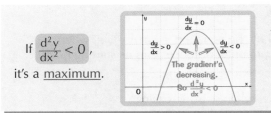

The gradient's decreasing. So $\frac{d^2y}{dx^2} < 0$

If $\frac{d^2y}{dx^2} > 0$, it's a <u>minimum</u>.

The gradient's increasing. So $\frac{d^2y}{dx^2} > 0$

But if $\frac{d^2y}{dx^2} = 0$, you can't tell what type of stationary point it is.

You've just found that $\frac{dy}{dx} = 6x^2 - 6x - 12$.

Stick in the x-coordinates of the stationary points.

So differentiating again gives $\frac{d^2y}{dx^2} = 12x - 6$.

At $x = -1$, $\frac{d^2y}{dx^2} = -18$, which is <u>negative</u> — so $x = -1$ is a <u>maximum</u>.

And at $x = 2$, $\frac{d^2y}{dx^2} = 18$, which is <u>positive</u> — so $x = 2$ is a <u>minimum</u>.

And since a cubic graph (where the coefficient of x^3 is <u>positive</u>) goes from <u>bottom-left to top-right</u>...

...you can draw a rough sketch of the graph, even though the roots would be hard to find.

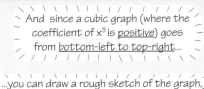

Stationary Points

1) **Find stationary points by solving** $\frac{dy}{dx} = 0$.

2) **Differentiate again to decide whether this is a maximum or a minimum.**

3) **If** $\frac{d^2y}{dx^2} < 0$ **— it's a maximum.**

   If $\frac{d^2y}{dx^2} > 0$ **— it's a minimum.**

An anagram of differentiation is "Perfect Insomnia Cure"...

No joke, is it — this differentiation business — but it's a dead important topic in maths. It's so important to know how to find whether a stationary point is a max or a min — but it can get a bit confusing. Try remembering MINMAX — which is short for 'MINUS means a MAXIMUM'. Or make up some other clever way to remember what means what.

Increasing and Decreasing Functions

Differentiation is all about finding gradients. Which means that you can find out where a graph is going up...
...and where it's going down. Lovely.

Find out if a function is Increasing or Decreasing

You can use differentiation to work out exactly where a function is <u>increasing</u> or <u>decreasing</u> — and how quickly.

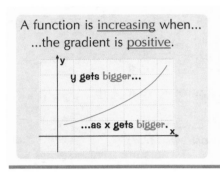

A function is <u>increasing</u> when...
...the gradient is <u>positive</u>.

y gets bigger...

...as x gets bigger.

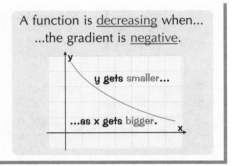

A function is <u>decreasing</u> when...
...the gradient is <u>negative</u>.

y gets smaller...

...as x gets bigger.

And there's more...

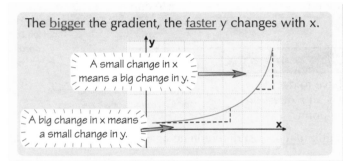

The <u>bigger</u> the gradient, the <u>faster</u> y changes with x.

A small change in x means a big change in y.

A big change in x means a small change in y.

Differentiation and Gradients

<u>Differentiate the equation</u> of the curve to find an expression for its gradient.

1) An increasing function has a <u>positive</u> gradient.

2) A decreasing function has a <u>negative</u> gradient.

Example: The path of a ball thrown through the air is described by the equation $y = 10x - 5x^2$, where y is the height of the ball above the ground and x is the horizontal distance from its starting point. Find where the height of the ball is <u>increasing</u> and where it's <u>decreasing</u>.

You have the equation for the path of the ball, and you need to know where y is increasing and where it's decreasing. That makes this a question about <u>gradients</u> — so <u>differentiate</u>.

$$y = 10x - 5x^2 \text{ so } \frac{dy}{dx} = 10 - 10x$$

This is an <u>increasing</u> function when: $10 - 10x > 0$, i.e. when $x < 1$, so the ball's height is increasing for $0 \le x < 1$.

And it's is a <u>decreasing</u> function when: $10 - 10x < 0$, i.e. when $x > 1$, so its height decreases for $x > 1$ (until it lands).

Just to check: The gradient of the ball's path is given by $\frac{dy}{dx} = 10 - 10x$.

There's a turning point (i.e. the ball's flight levels out) when $x = 1$.

Differentiating again gives $\frac{d^2y}{dx^2} = -10$, which is <u>negative</u> — and so the turning point is a <u>maximum</u>.

This is what you'd expect — the ball goes <u>up then down</u>, not the other way round.

Help me Differentiation — You're my only hope...

There's not much hard maths on this page — but there are a couple of very important ideas that you need to get your head round pretty darn soon. Understanding that differentiating gives the gradient of the graph is more important than washing regularly — AND THAT'S IMPORTANT. The other thing on the page — that you can tell whether a function is getting bigger or smaller by looking at the derivative — is also vital. Sometimes the examiners ask you to find where a function is increasing or decreasing — so you'd just have to find where the derivative was positive or negative.

Curve Sketching

You'll even be asked to do some drawing in the exam... but don't get too excited — it's just drawing graphs... great.

Find where the curve crosses the Axes

Sketch the graph of $f(x) = \frac{x^2}{2} - 2\sqrt{x}$, for $x \geq 0$.

The curve crosses the y-axis when x = 0 — so put x = 0 in the expression for y.

When x = 0, f(x) = 0 — and so the curve goes through the <u>origin</u>.

The curve crosses the x-axis when f(x) = 0. So solve

$$\frac{x^2}{2} - 2\sqrt{x} = 0$$

$$\Rightarrow x^2 = 4\sqrt{x} = 4x^{\frac{1}{2}}$$

Dividing both sides by $x^{\frac{1}{2}}$.

$$\Rightarrow x^{\frac{3}{2}} = 4$$

$$\Rightarrow x = 4^{\frac{2}{3}} = \sqrt[3]{4^2} = \sqrt[3]{16} \approx 2.5$$

And so the curve crosses the x-axis when x ≈ 2.5 .

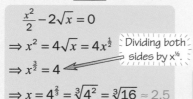

...Differentiate to find Gradient info...

Differentiating the function gives...

$$f(x) = \frac{1}{2}x^2 - 2x^{\frac{1}{2}}$$

$$\Rightarrow f'(x) = \frac{1}{2}(2x) - 2\left(\frac{1}{2}x^{-\frac{1}{2}}\right) = x - x^{-\frac{1}{2}} = x - \frac{1}{\sqrt{x}}$$

Using the derivative — you can find stationary points and tell when the graph goes 'uphill' and 'downhill'.

1) So there's a <u>stationary point</u> when...

$$x - \frac{1}{\sqrt{x}} = 0$$

$$\Rightarrow x = \frac{1}{\sqrt{x}}$$

$$\Rightarrow x^{\frac{3}{2}} = 1 \Rightarrow x = 1$$

And at x = 1, $f(x) = \frac{1}{2} - 2 = -\frac{3}{2}$.

2) The gradient's <u>negative</u> when...

$$x - \frac{1}{\sqrt{x}} < 0$$

$$\Rightarrow x < \frac{1}{\sqrt{x}}$$

$$\Rightarrow x^{\frac{3}{2}} < 1 \Rightarrow x < 1$$

So the function <u>decreases</u> when $0 \leq x < 1$...

3) The gradient's <u>positive</u> when...

$$x - \frac{1}{\sqrt{x}} > 0$$

$$\Rightarrow x > 1$$

...and <u>increases</u> for x > 1.

This is the quickest way to check if something's a max or a min.

You could check that x = 1 is a <u>minimum</u> by differentiating again.

$$f''(x) = 1 - \left(-\frac{1}{2}x^{-\frac{3}{2}}\right) = 1 + \frac{1}{2\sqrt{x^3}}$$

This is <u>positive</u> when x = 1, and so this is definitely a minimum.

...and find out what happens when x gets Big

You can also try and decide what happens as x gets very <u>big</u> — in both the positive and negative directions.

Factorise f(x) by taking the <u>biggest</u> power outside the brackets...

$$\frac{x^2}{2} - 2\sqrt{x} = x^2\left(\frac{1}{2} - 2x^{-\frac{3}{2}}\right) = x^2\left(\frac{1}{2} - \frac{2}{x^{\frac{3}{2}}}\right)$$

As x gets large, this bit disappears — and the bit in brackets gets closer to $\frac{1}{2}$.

And so as x gets larger, f(x) gets closer and closer to $\frac{1}{2}x^2$ — and this just keeps growing and growing.

And the graph looks like this...

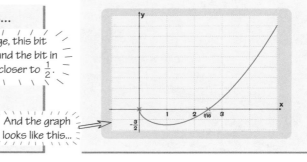

Curve-sketching's important — but don't take my word for it...

Curve-sketching — an underrated skill, in my opinion. As Shakespeare once wrote, 'Those who can do fab sketches of graphs and stuff are likely to get pretty good grades in maths exams, no word of a lie'. Well, he probably would've written something like that if he was into maths. And he would've written it because graphs are helpful when you're trying to work out what a question's all about — and once you know that, you can decide the best way forward. And if you don't believe me, remember the saying of the ancient Roman Emperor Julius Caesar, 'If in doubt, draw a graph'.

Section Five Revision Questions

Well, I'll be... another section all over and done with. And there's some mighty useful stuff in that last section too. Now drawing graphs is something that it's always handy to be able to do, which makes differentiation a handy skill to have (as well as being something you have to be comfortable with by the time you step in to do your Core 2 exam). The questions below will help you get to grips with it all. Try them and see how you get on. If you get all the answers right, then well done... go get yourself a pie. But if you get any wrong, read the section again, work out where you went wrong, and then try the questions again.

1) a) Write down what a stationary point is.

   b) Find the stationary points of the graph of $y = x^3 - 6x^2 - 63x + 21$.

2) How can you decide whether a stationary point is a maximum or a minimum?

3) A bit fiddly this — you won't like it — so just make sure you can do it.
   Find the stationary points of the function $y = x^3 + \frac{3}{x}$.
   Decide whether each stationary point is a minimum or a maximum.

4) Not so easy — it involves inequalities: find when these two functions are increasing and decreasing:
   a) $y = 6(x + 2)(x - 3)$,

   b) $y = \frac{1}{x^2}$.

Integration

Some integrals have <u>limits</u> (i.e. little numbers) next to the integral sign. You integrate them in exactly the same way — but you <u>don't</u> need a constant of integration. Much easier. And scrummier and yummier too.

A *Definite Integral* finds the *Area Under a Curve*

This definite integral tells you the <u>area</u> between the graph of $y = x^3$ and the x-axis between x = –2 and x = 2:

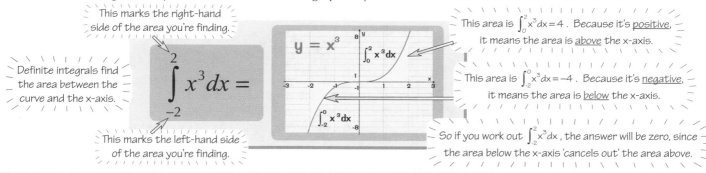

This marks the right-hand side of the area you're finding.

Definite integrals find the area between the curve and the x-axis.

This marks the left-hand side of the area you're finding.

$$\int_{-2}^{2} x^3 dx =$$

This area is $\int_{0}^{2} x^3 dx = 4$. Because it's <u>positive</u>, it means the area is <u>above</u> the x-axis.

This area is $\int_{-2}^{0} x^3 dx = -4$. Because it's <u>negative</u>, it means the area is <u>below</u> the x-axis.

So if you work out $\int_{-2}^{2} x^3 dx$, the answer will be zero, since the area below the x-axis 'cancels out' the area above.

Do the integration in the same way — then use the *Limits*

Finding a definite integral isn't really any harder than an indefinite one — there's just an <u>extra</u> stage you have to do. After you've integrated the function you have to work out the value of this new function by sticking in the <u>limits</u>.

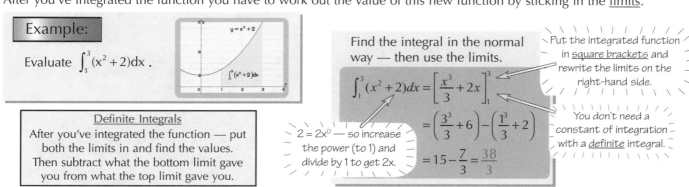

Example:

Evaluate $\int_{1}^{3} (x^2 + 2) dx$.

Definite Integrals
After you've integrated the function — put both the limits in and find the values. Then subtract what the bottom limit gave you from what the top limit gave you.

Find the integral in the normal way — then use the limits.

Put the integrated function in <u>square brackets</u> and rewrite the limits on the right-hand side.

$$\int_{1}^{3} (x^2 + 2) dx = \left[\frac{x^3}{3} + 2x \right]_{1}^{3}$$

$2 = 2x^0$ — so increase the power (to 1) and divide by 1 to get 2x.

$$= \left(\frac{3^3}{3} + 6 \right) - \left(\frac{1^3}{3} + 2 \right)$$

$$= 15 - \frac{7}{3} = \frac{38}{3}$$

You don't need a constant of integration with a <u>definite</u> integral.

Integrate *'to Infinity'* with the ∞ *(infinity) sign*

And you can integrate all the way to <u>infinity</u> as well. Just use the ∞ symbol as your upper limit. Or use –∞ as your lower limit if you want to integrate to '<u>minus infinity</u>'.

Example: Find the area under the curve $y = \frac{15}{x^2} - \frac{30}{x^3}$ for $x \geq 2$.

For this, you need to integrate from x = 2 up to infinity (∞).

Use this sign as a limit to integrate 'to infinity'.

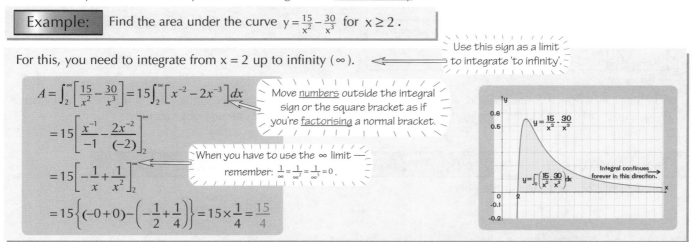

$$A = \int_{2}^{\infty} \left[\frac{15}{x^2} - \frac{30}{x^3} \right] = 15 \int_{2}^{\infty} \left[x^{-2} - 2x^{-3} \right] dx$$

Move <u>numbers</u> outside the integral sign or the square bracket as if you're <u>factorising</u> a normal bracket.

$$= 15 \left[\frac{x^{-1}}{-1} - \frac{2x^{-2}}{(-2)} \right]_{2}^{\infty}$$

$$= 15 \left[-\frac{1}{x} + \frac{1}{x^2} \right]_{2}^{\infty}$$

When you have to use the ∞ limit — remember: $\frac{1}{\infty} = \frac{1}{\infty^2} = \frac{1}{\infty^3} = 0$.

$$= 15 \left\{ (-0 + 0) - \left(-\frac{1}{2} + \frac{1}{4} \right) \right\} = 15 \times \frac{1}{4} = \frac{15}{4}$$

My hobbies? Well I'm really inte grating. Especially carrots.

It's still integration — but this time you're putting two numbers into an equation afterwards. So although this may not be the wild and crazy fun-packed time your teachers promised you when they were trying to persuade you to take AS maths, you've got to admit that a lot of this stuff is pretty similar — and if you can do one bit, you can use that to do quite a few other bits too. Maths is like that. But I admit it's probably not as much fun as a big banana-and-toffee cake.

The Trapezium Rule

Sometimes <u>integrals</u> can be just <u>too hard</u> to do using the normal methods — then you need to know other ways to solve them. That's where the <u>Trapezium Rule</u> comes in.

The **Trapezium Rule** is Used to Find the **Approximate Area** Under a Curve

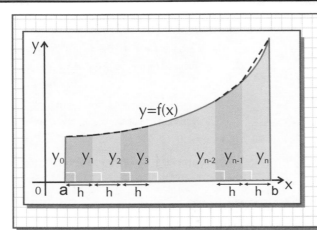

The area represented by $\int_a^b y\,dx$ is approximately:

$$\int_a^b y\,dx \approx \frac{h}{2}[y_0 + 2(y_1 + y_2 + ... + y_{n-1}) + y_n]$$

where **n** is the number of strips or intervals and **h** is the width of each strip.

You can find the width of each strip using $h = \frac{(b-a)}{n}$

$y_0, y_1, y_2, ... , y_n$ are the heights of the sides of the trapeziums — you get these by putting the x-values into the curve.

So basically the formula for approximating $\int_a^b y\,dx$ works like this:

'Add the first and last heights $(y_0 + y_n)$ and add this to <u>twice</u> all the other heights added up — then multiply by $\frac{h}{2}$.'

Example: Find an approximate value for $\int_0^2 \sqrt{4-x^2}\,dx$ using 4 strips. Give your answer to 4 s.f.

Start by working out the width of each strip: $h = \frac{(b-a)}{n} = \frac{(2-0)}{4} = 0.5$

This means the x-values are $x_0 = 0$, $x_1 = 0.5$, $x_2 = 1$, $x_3 = 1.5$ and $x_4 = 2$ (the question specifies 4 strips, so n = 4). Set up a table and work out the y-values or heights using the equation in the integral.

x	$y = \sqrt{4-x^2}$
$x_0 = 0$	$y_0 = \sqrt{4-0^2} = 2$
$x_1 = 0.5$	$y_1 = \sqrt{4-0.5^2} = \sqrt{3.75} = 1.936491673$
$x_2 = 1.0$	$y_2 = \sqrt{4-1.0^2} = \sqrt{3} = 1.732050808$
$x_3 = 1.5$	$y_3 = \sqrt{4-1.5^2} = \sqrt{1.75} = 1.322875656$
$x_4 = 2.0$	$y_4 = \sqrt{4-2.0^2} = 0$

Now put all the y-values into the formula with h and n:

$$\int_a^b y\,dx \approx \frac{0.5}{2}[2 + 2(1.936491673 + 1.732050808 + 1.322875656) + 0]$$
$$\approx 0.25[2 + 2 \times 4.991418137]$$
$$\approx 2.996 \text{ to 4 s.f.}$$

Watch out — if they ask you to work out a question with 5 y-values (or '<u>ordinates</u>') then this is the <u>same</u> as 4 strips. The x-values usually go up in <u>nice jumps</u> — if they don't then <u>check</u> your calculations carefully.

The Approximation might be an **Overestimate** or an **Underestimate**

It all depends on the shape of the curve...

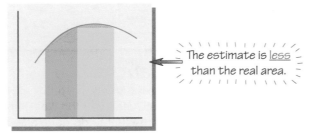

The estimate is <u>less</u> than the real area.

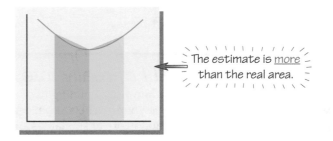

The estimate is <u>more</u> than the real area.

The Trapezium Rule

These are usually popular questions with examiners — as long as you're careful there are <u>plenty of marks</u> to be had.

The **Trapezium Rule** is in the **Formula Booklet**

...so don't try any heroics — always <u>look it up</u> and use it with these questions.

Example: Use the trapezium rule with 7 ordinates to find an approximation to $\int_1^{2.2} 2\log_{10} x \, dx$

Remember, <u>7 ordinates</u> means <u>6 strips</u> — so n = 6.

Calculate the width of the strips: $h = \dfrac{(b-a)}{n} = \dfrac{(2.2-1)}{6} = 0.2$

Set up a table and work out the y-values using $y = 2\log_{10} x$:

x	$y = 2 \log_{10} x$
x_0=1.0	y_0= 2 log$_{10}$ 1 = 0
x_1=1.2	y_1= 2 log$_{10}$ 1.2 = 0.15836
x_2=1.4	y_2=0.29226
x_3=1.6	y_3=0.40824
x_4=1.8	y_4=0.51055
x_5=2.0	y_5=0.60206
x_6=2.2	y_6=0.68485

$y_6 = 2 \log_{10} b = 0.68485$

Putting all these values in the formula gives:

$\int_a^b y\,dx \approx \dfrac{0.2}{2} [0+2(0.15836+0.29226+0.40824+0.51055+0.60206)+0.68485]$

$\approx 0.1\times[0.68485+2\times1.97147]$

≈ 0.462777

≈ 0.463 *to 3 d.p.*

Example: Use the trapezium rule with 8 intervals to find an approximation to $\int_0^{\pi} \sin x \, dx$

Whenever you get a calculus question using <u>trig functions</u>, you <u>have</u> to use <u>radians</u>. You'll probably be given a limit with π in, which is a pretty good reminder.

There are 8 intervals, so n = 8.

Keep your x-values in terms of π.

Calculate the width of the strips: $h = \dfrac{(b-a)}{n} = \dfrac{(\pi-0)}{8} = \dfrac{\pi}{8}$

Set up a table and work out the y-values:

x	$y = \sin x$
x_0= 0	y_0= sin 0 = 0
$x_1 = \frac{\pi}{8}$	y_1= 0.38268
$x_2 = \frac{\pi}{4}$	y_2=0.70711
$x_3 = \frac{3\pi}{8}$	y_3=0.92388
$x_4 = \frac{\pi}{2}$	y_4=1
$x_5 = \frac{5\pi}{8}$	y_5=0.92388
$x_6 = \frac{3\pi}{4}$	y_6=0.70711
$x_7 = \frac{7\pi}{8}$	y_7= 0.38268
$x_8 = \pi$	y_8= 0

So, putting all this in the formula gives:

$\int_a^b y\,dx \approx \dfrac{1}{2}\cdot\dfrac{\pi}{8}[0+2(0.38268+0.70711+0.92388+1+0.92388+0.70711+0.38268)+0]$

$\approx \dfrac{\pi}{16}[2\times5.02734]$

≈ 1.9742

≈ 1.974 to 3 d.p.

Maths rhyming slang #3: Dribble and drool — Trapezium rule...

Take your time with Trapezium Rule questions — it's so easy to make a mistake with all those numbers flying around. Make a nice table showing all your ordinates (careful — this is always one more than the number of strips). Then add up y_1 to y_{n-1} and multiply the answer by 2. Add on y_0 and y_n. Finally, multiply what you've got so far by the width of a strip and <u>divide by 2</u>. It's a good idea to write down what you get after each stage, by the way — then if you press the wrong button (easily done) you'll be able to pick up from where you went wrong. They're not hard — just fiddly.

Areas Between Curves

With a bit of thought, you can use integration to find all kinds of areas — even ones that look quite tricky at first. The best way to work out what to do is draw a <u>picture</u>. Then it'll seem easier. I promise you it will.

Sometimes you have to **Add** integrals...

This looks pretty hard — until you draw a picture and see what it's all about.

Example: Find the area enclosed by the curves $y = x^2$, $y = (2 - x)^2$ and the x-axis.

Find out where the curves meet by <u>solving</u> $x^2 = (2-x)^2$. — they meet at x=1.

You have to find area A — but you'll need to <u>split</u> it into two smaller pieces.

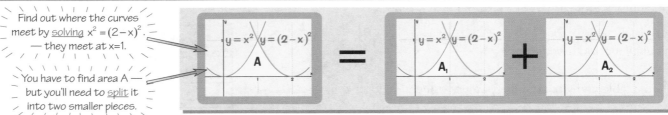

And it's pretty clear from the picture that you'll have to find the area in two lumps, A_1 and A_2.

The first area you need to find is A_1:

$$A_1 = \int_0^1 x^2 \, dx$$

$$= \left[\frac{x^3}{3} \right]_0^1$$

$$= \left(\frac{1}{3} - 0 \right) = \frac{1}{3}$$

The other area you need is A_2:

$$A_2 = \int_1^2 (2-x)^2 \, dx = \int_1^2 (4 - 4x + x^2) \, dx$$

$$= \left[4x - 2x^2 + \frac{x^3}{3} \right]_1^2$$

$$= \left(8 - 8 + \frac{8}{3} \right) - \left(4 - 2 + \frac{1}{3} \right)$$

$$= \frac{8}{3} - \frac{7}{3} = \frac{1}{3}$$

And the area the question actually asks for is $A_1 + A_2$. This is

$$A = A_1 + A_2$$

$$= \frac{1}{3} + \frac{1}{3} = \frac{2}{3}$$

If you spot that the area A is <u>symmetrical</u> about $x = 1$, you can save yourself some work by calculating half the area and then doubling it: $A = 2\int_0^1 x^2 dx$

...sometimes you have to **Subtract** them

Again, it's best to look at the <u>pictures</u> to work out exactly what you need to do.

Example: Find the area enclosed by the curves $y = x^2 + 1$ and $y = 9 - x^2$.

Solve $x^2 + 1 = 9 - x^2$ to find where the curves meet.
$x^2 + 1 = 9 - x^2 \Rightarrow 2x^2 = 8$
$\Rightarrow x^2 = 4$
$\Rightarrow x = \pm 2$

So you'll have to integrate between −2 and 2.

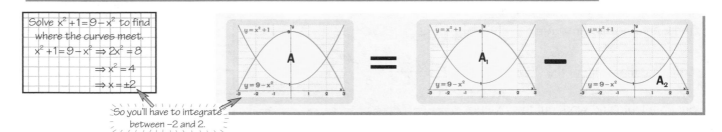

The area under the green curve A_1 is:

$$A_1 = \int_{-2}^2 (9 - x^2) \, dx$$

$$= \left[9x - \frac{x^3}{3} \right]_{-2}^2$$

$$= \left(18 - \frac{2^3}{3} \right) - \left(-18 - \frac{(-2)^3}{3} \right)$$

$$= \left(18 - \frac{8}{3} \right) - \left(-18 - \left(-\frac{8}{3} \right) \right) = \frac{46}{3} - \left(-\frac{46}{3} \right) = \frac{92}{3}$$

The area under the red curve is:

$$A_2 = \int_{-2}^2 (x^2 + 1) \, dx$$

$$= \left[\frac{x^3}{3} + x \right]_{-2}^2$$

$$= \left(\frac{2^3}{3} + 2 \right) - \left(\frac{(-2)^3}{3} + (-2) \right)$$

$$= \left(\frac{8}{3} + 2 \right) - \left(-\frac{8}{3} - 2 \right) = \frac{28}{3}$$

And the area you need is the difference between these:

$$A = A_1 - A_2$$

$$= \frac{92}{3} - \frac{28}{3} = \frac{64}{3}$$

Instead of integrating before subtracting — you could try 'subtracting the lines', and then integrating. This last area A is also:

$$A = \int_{-2}^2 \{ (9 - x^2) - (x^2 + 1) \} dx$$

And so, our hero integrates the area between two curves, and saves the day...

That's the basic idea of finding the area enclosed by two curves and lines — draw a picture and then break the area down into smaller, easier chunks. And it's always a good idea to keep an eye out for anything symmetrical that could save you a bit of work — like in the first example. Questions like this aren't hard — but they can sometimes take a long time. Great.

Section Six Revision Questions

They think it's all over...

1) How can you tell whether an integral is a definite one or an indefinite one?
   (It's easy really — it just sounds difficult.)

2) What does a definite integral represent on a graph?

3) Evaluate the following definite integrals:

   a) $\int_0^1 (4x^3 + 3x^2 + 2x + 1)dx$, *b)* $\int_1^2 \left(\frac{8}{x^5} + \frac{3}{\sqrt{x}}\right)dx$, *c)* $\int_1^6 \frac{3}{x^2}dx$.

4) Evaluate: a) $\int_{-3}^3 (9 - x^2)dx$, b) $\int_1^\infty \frac{3}{x^2}dx$.

   Sketch the areas represented by these integrals.

5) Find area A in the diagram below:

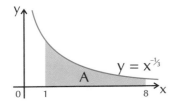

6) Use the trapezium rule with n intervals to estimate:

   a) $\int_0^3 (9 - x^2)^{\frac{1}{2}}dx$ with $n = 3$

   b) $\int_{0.2}^{1.2} x^{x^2}dx$; $n = 5$

7) Find the yellow area in each of these graphs:

   a)
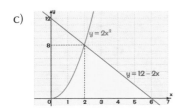

   b)
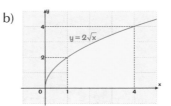

   c)
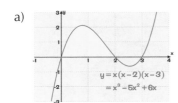

   d)
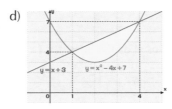

...it is now.

General Certificate of Education
Advanced Subsidiary (AS) and Advanced Level

Core 2 Mathematics — Practice Exam One

Give non-exact numerical answers correct to 3 significant figures, unless a
different degree of accuracy is specified in the question or is clearly appropriate.

---

1 **(i)** By sketching the graph of $y = \tan 2t$ for a suitable range of t, determine the number of solutions to the equation

$$\tan 2t = k \qquad \text{in the range } 0° \le t < 360°, \text{where k is any number.}$$ [3]

    (ii) Solve the equation $\sin 2t = \sqrt{2} \cos 2t$, giving all the solutions in the range $0° \le t < 360°$. [3]

---

2 **(i)** Write down the value of $\log_3 3$ [1]

    (ii) Given that $\log_a \chi = \log_a 4 + 3 \log_a 2$ show that $\chi = 32$ [2]

---

3 **(i)** Sketch the curve $y = \dfrac{1}{x^2}$ for $x > 0$. [1]

    (ii) Show that $\displaystyle\int_1^\infty \dfrac{1}{x^2}\, dx = 1.$ [2]

    (iii) Find f(x), where $y = $ f(x) is the equation of the tangent to the graph of $y = \dfrac{1}{x^2}$ at the point where $x = 1$. [2]

    (iv) Find k < 1 such that $\displaystyle\int_0^k f(x)\,dx = 1$, where f(x) is the function found in part (iii).

    Give your answer using surds. [3]

---

4 The derivative of a function is given by $\dfrac{dy}{dx} = \dfrac{1}{2}x^2 - \dfrac{3}{\sqrt{x}}$

    (i) Find an expression for y if the graph of y against x is to pass through the point $\left(1, \tfrac{1}{6}\right)$. [4]

    (ii) Evaluate $\displaystyle\int_0^1 y\,dx.$ [4]

---

5 **(i)** Sketch the curve $y = (x-2)(x-4)$ and the line $y = 2x - 4$ on the same set of axes, clearly
marking the coordinates of the points of intersection. [3]

    (ii) Evaluate the integral $\displaystyle\int_2^4 (x-2)(x-4)\,dx$. [3]

    (iii) Hence or otherwise, show the total area enclosed by the lines $y = (x-2)(x-4)$ and $y = 2x-4$ is $\dfrac{32}{3}$. [4]

6   The diagram below shows a sector of a circle of radius r cm and angle $120°$.

The length of the arc of the sector is 40 cm.

(i)   Write $120°$ in radians. [1]

(ii)   Show that $r \approx 19.1$ cm. [2]

(iii) Find the area of the sector to the nearest square centimetre. [3]

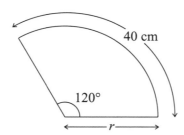

7   **(i)**   Rewrite the following expression in the form $f(x) = 0$, where $f(x)$ is of the form $f(x) = ax^3 + bx^2 + cx + d$.
$$(x-1)(x^2 + x + 1) = 2x^2 - 17$$ [2]

(ii)   Show that $(x+2)$ is a factor of $f(x)$. [3]

(iii) Using your answer to part (ii), factorise $f(x)$ as the product of a linear factor and a quadratic factor. [3]

(iv) By completing the square, or otherwise, show that $f(x) = 0$ has no other roots, and sketch the graph of $f(x)$. [2]

(v)   Divide the polynomial $x^3 - 2x^2 + 3x - 3$ by $(x - 1)$, showing both the quotient and remainder. [4]

8   The diagram shows a circle. A $(2, 1)$ and B $(0, -5)$ lie on the circle and AB is a diameter.

C $(4, -1)$ is also on the circle.

(i)   Find the centre and radius of the circle. [4]

(ii)   Show that the equation of the circle can be written in the form:
$$x^2 + y^2 - 2x + 4y - 5 = 0$$ [3]

(iii) The tangent at A and the normal at C cross at D.

Find the coordinates of D. [7]

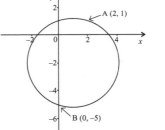

9   For the series with second term -2 and common ratio -½, find:

(i)   the first seven terms [3]

(ii)   the sum to infinity [3]

Paper 1 Q1 — Trigonometry

1 **(i)** By sketching the graph of $y = \tan 2t$ for a suitable range of t, determine the number of solutions to the equation

$\tan 2t = k$ in the range $0° \le t < 360°$, where k is any number. [3]

(ii) Solve the equation $\sin 2t = \sqrt{2}\cos 2t$, giving all the solutions in the range $0° \le t < 360°$. [3]

(i) A *Squashed up* tan graph

' Sketch the graph of $y = \tan 2t$ '

Hmm, that looks suspiciously like $y = \tan x$ to me. Just write t instead of x, and draw the graph <u>twice as squashed</u>.

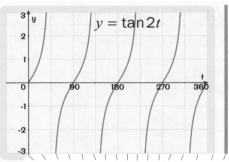

$$y = \tan 2t$$

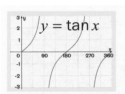

The '2' means the period is <u>twice as short</u> (or half as long) — 90°, not 180°. A common mistake is to draw the graph twice as long.

It says sketch the graph for a <u>suitable</u> range of t — so sketch the range $0° \le t < 360°$, since that's the range you have to find solutions in.

Use your calculator to check the points when the graph should be zero and infinity. If you get an answer you don't expect — stop and think.

Think of 'y=k' — a horizontal line

Now all this is basically asking is, 'how many times does the graph cross the line y = k?'.

Now, y = k is just a horizontal line, and it doesn't make any difference exactly where — you can see from the graph that y = tan 2t will always cross it 4 times in the range $0° \le t < 360°$.

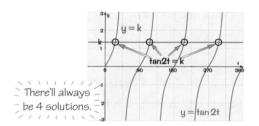

There'll always be 4 solutions.

(ii) It's the *Same Equation*, in disguise

' Solve the equation $\sin 2t = \sqrt{2}\cos 2t$.'

It looks a bit tricky at first, until you realise that if you divide by cos 2t, you get $\tan 2t = \sqrt{2}$ (as $\tan = \frac{sin}{cos}$).

The first part of the question was to prepare you to answer this bit.

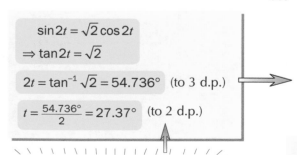

$\sin 2t = \sqrt{2}\cos 2t$

$\Rightarrow \tan 2t = \sqrt{2}$

$2t = \tan^{-1}\sqrt{2} = 54.736°$ (to 3 d.p.)

$t = \frac{54.736°}{2} = 27.37°$ (to 2 d.p.)

Always use more decimal places in your workings than you want for your answer — that way you won't make a rounding error.

Whatever you do — <u>don't</u> stop there. That's only one solution, but you know from the first part that there's got to be <u>four</u>.

It's not as hard as it sounds, though. The graph <u>repeats</u> itself every 90°, so all you've got to do is add on 90° three times to get the other answers:

So the four solutions to the equation are...

27.37°, 117.37°, 207.37° and 297.37°.

Trig or happy? Trig or treat? Trig or finger? It's just trig-tastic...

There are two important things to be learnt here: Firstly, make sure you know all those standard graphs — then the first part of the question really should be easy. And secondly, if you're stuck on the last part of the question, always look at the earlier bits for hints — it's a pretty safe bet there's going to be some kind of connection that'll help.

Paper 1 Q2 — Logs

> **2 (i)** Write down the value of $\log_3 3$ [1]
>
> **(ii)** Given that $\log_a \chi = \log_a 4 + 3\log_a 2$ show that $\chi = 32$ [2]

(i) $Log_3 3$ — Some Marks are a *Giveaway*...

'Write down the value of $\log_3 3$'

This isn't a trick question — they're just checking you know what log means.
From page 17 you should (hopefully) remember that:

> $\log_a b = c$ means the same as $a^c = b$
>
> That means that $\log_a a = 1$ and $\log_a 1 = 0$

From that bit, it's pretty easy to work out that $\log_3 3 = 1$.

(In other words, $3^1 = 3$, which you should also know from the laws of indices that you did in Core 1.)

(ii) For *Log Equations* you need to *Learn* the *Log Laws*

You'll get the first mark for showing you can use one of the <u>laws of logarithms</u> and the other for successfully getting $\chi = 32$.

OK, let's get stuck in...

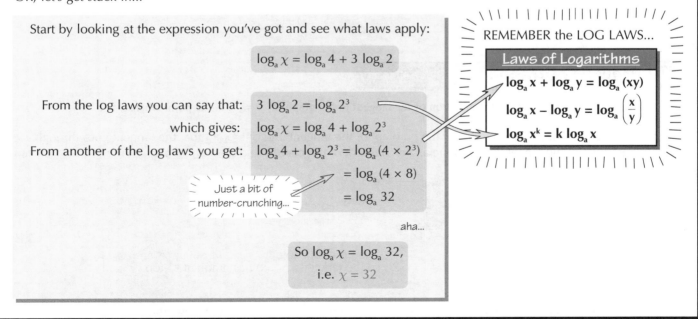

Start by looking at the expression you've got and see what laws apply:

$$\log_a \chi = \log_a 4 + 3\log_a 2$$

From the log laws you can say that: $\quad 3\log_a 2 = \log_a 2^3$

which gives: $\quad \log_a \chi = \log_a 4 + \log_a 2^3$

From another of the log laws you get: $\quad \log_a 4 + \log_a 2^3 = \log_a (4 \times 2^3)$

$$= \log_a (4 \times 8)$$

$$= \log_a 32$$

Just a bit of number-crunching...

aha...

So $\log_a \chi = \log_a 32$,

i.e. $\chi = 32$

REMEMBER the LOG LAWS...

Laws of Logarithms

$$\log_a x + \log_a y = \log_a (xy)$$

$$\log_a x - \log_a y = \log_a \left(\frac{x}{y}\right)$$

$$\log_a x^k = k\log_a x$$

AS Maths — (nearly) as easy as falling off a log...

OK, logs aren't the easiest thing in the world, but this is a straightforward question so quit complaining. The best way to prepare for an exam question on logs is to *LEARN THE LOG LAWS*. Whether or not you can apply those laws is kind of irrelevant if you don't know them in the first place. So I repeat: *LEARN THE LOG LAWS*... and *then* practise using them.

Paper 1 Q3 — Integration and Tangents

3 **(i)** Sketch the curve $y = \frac{1}{x^2}$ for $x > 0$. [1]

(ii) Show that $\int_{1}^{\infty} \frac{1}{x^2} dx = 1$. [2]

(iii) Find f(x), where $y = $ f(x) is the equation of the tangent to the graph of $y = \frac{1}{x^2}$ at the point where $x = 1$. [2]

(iv) Find $k < 1$ such that $\int_{0}^{k} f(x) dx = 1$, where f(x) is the function found in part (iii). Give your answer using surds. [3]

(i) It's only a **Sketch** — but get it right

'Sketch the curve $y = \frac{1}{x^2}$ for $x > 0$.'

You need to know what this sort of graph looks like — so this should be <u>easy</u> marks. (They're covered on page 29.)

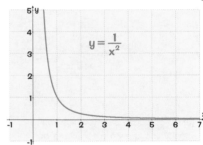

But <u>if</u> you can't remember, you can work it out easily enough:

You need to sketch y for x > 0. So just look at what happens to y:
a) as x gets <u>close to zero</u>   and   b) as x <u>gets very big</u>.

a)   As x gets very large (i.e. tends towards infinity), 1 / x² gets very small.

b)   As x tends towards zero, 1/x² tends towards infinity.

(You should use this to check your graph anyway.)

(ii) An integral to **Infinity** — but at least they tell you the answer

'Show that $\int_{1}^{\infty} \frac{1}{x^2} dx = 1$.'

The integration isn't too bad but take care with the limits, and don't 'lose' any <u>minus signs</u>.

Integrating between 1 and infinity gives...

$$\int_{1}^{\infty} \frac{1}{x^2} dx = \int_{1}^{\infty} x^{-2} dx$$

Increase the power by 1 (to make −1) and then divide by −1.

$$= \left[\frac{x^{-1}}{-1} \right]_{1}^{\infty} = \left[-\frac{1}{x} \right]_{1}^{\infty}$$

$$= (-0) - \left(-\frac{1}{1} \right) = 1$$

Since $\frac{1}{\infty} = 0$.

(iii) Now find a **Tangent** to the curve

'Find f(x), where $y = $ f(x) is the equation of the tangent to the graph of $y = \frac{1}{x^2}$ at the point where $x = 1$.'

To find a tangent, you just need to know the <u>gradient</u> of the tangent and one <u>point</u> that the tangent goes through. But the gradient of the tangent is the same as the gradient of the curve — so <u>differentiate</u>.

$$y = \frac{1}{x^2} = x^{-2}$$

$$\Rightarrow \frac{dy}{dx} = -2x^{-3} = -\frac{2}{x^3}$$

This is the gradient of the curve $y = \frac{1}{x^2}$.

The gradient of the curve at x = 1 is $-\frac{2}{1^3} = -2$, so the tangent y = f(x) has gradient **-2** too.

Since y = f(x) is a straight line, write its equation as y = mx + c, where <u>m</u> is the gradient.

$$y = -2x + c$$

Now you just need to find c.

Paper 1 Q3 — Integration and Tangents

To find c, you need to use the fact that the tangent <u>touches</u> the curve at the point (1, 1).

The tangent goes through the point (1, 1), and so

$$y = -2x + c$$
$$\Rightarrow 1 = (-2 \times 1) + c = -2 + c$$

Since y = 1.

$$\Rightarrow c = 3$$

And this means the equation of the tangent is...

$$y = -2x + 3$$

and so $f(x) = 3 - 2x$

This means the same but looks a bit neater.

(iv) Another **Integration** — but this time you have to find the **Upper Limit**

'Find $k < 1$ such that $\int_0^k f(x)\,dx = 1$, where f(x) is the function found in part (iii). Give your answer using surds.'

Now you need to find a value less than 1 for k. Just do the integration <u>normally</u> but write k instead of a number...

$$\int_0^k f(x)\,dx = \int_0^k (3 - 2x)\,dx$$
$$= \left[3x - x^2\right]_0^k$$
$$= (3k - k^2) - 0$$
$$= 3k - k^2$$

Don't do anything differently just because you have k instead of a number.

This integral has to be equal to 1, so you need to solve

$$3k - k^2 = 1$$
$$\Rightarrow k^2 - 3k + 1 = 0$$

Since the question asks you to find k such that $\int_0^k f(x)\,dx = 1$.

This is a <u>quadratic</u>, and the question mentions <u>surds</u> — so it doesn't look as though it's going to factorise.

Using the quadratic formula...

$$k = \frac{3 \pm \sqrt{(-3)^2 - 4 \times 1 \times 1}}{2 \times 1}$$
$$= \frac{3 \pm \sqrt{5}}{2}$$

There are two possible values for k here — you have to decide which one you need.

The question says k has to be less than 1 — so use a calculator to work out which of these possible values you need.

$$\frac{3 + \sqrt{5}}{2} = 2.618 \qquad \frac{3 - \sqrt{5}}{2} = 0.382$$

So this is the one you need.

The question asks you to give your final answer in <u>surds</u>, though.

And so the required value of k is

$$k = \frac{3 - \sqrt{5}}{2}$$

---

f(x)? Can't they think of a better name? Like Trevor. Yes — Trevor The Tangent

Phew. I mean, really... phew. What a stinker. Let's break it down. <u>Part i)</u> — you really should know what graphs like y = kxⁿ look like (see page 26) — but you can work it out anyway if you just choose a few values for x, work out the value of y there, and then plot the points. <u>Part ii)</u> It's just a limit integration. Don't be put off by the ∞ — it only comes into play when you're substituting in the limits at the end (where you have to use the fact that 1/∞ is 0). <u>Part iii)</u> is okay really. Nuff said. <u>Part iv)</u> — well it's long. And you get two answers at the end when you might only expect one. When this happens, have another look at the question and see if you can get rid of one somehow. Here, they only want a value of k that's less than 1 — so you can get rid of the one that's bigger than 1. Obvious, eh? — but easily forgotten in the exam.

Paper 1 Q4 — Calculus

4   The derivative of a function is given by

$$\frac{dy}{dx} = \frac{1}{2}x^2 - \frac{3}{\sqrt{x}}$$

(i)   Find an expression for y if the graph of y against x is to pass through the point $\left(1, \frac{1}{6}\right)$.   [4]

(ii)   Evaluate $\int_0^1 y\,dx$.   [4]

(i) You just *Integrate* it...

'Find an expression for *y* if the graph of *y* against *x* is to pass through the point $\left(1, \frac{1}{6}\right)$.'

To get an expression for y from an expression for $\frac{dy}{dx}$, you just <u>integrate</u>.

(Integration is the opposite of differentiation, remember.)

Before you can integrate it, you have to rewrite $\frac{dy}{dx}$ as <u>powers of x</u>...

$$\frac{dy}{dx} = \frac{1}{2}x^2 - \frac{3}{\sqrt{x}}$$
$$= \frac{1}{2}x^2 - 3x^{-\frac{1}{2}}$$

Rewrite that square root as a power of x.

Now write both sides as integrals...

Just put an integral sign and 'dx' around both sides for now.

$$\int \frac{dy}{dx}\,dx = \int\left(\frac{1}{2}x^2 - 3x^{-\frac{1}{2}}\right)dx$$

Integrate the left-hand side and you get y.
And you can rewrite the right-hand side to break it into smaller bits...

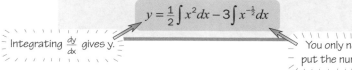

$$y = \frac{1}{2}\int x^2\,dx - 3\int x^{-\frac{1}{2}}\,dx$$

Integrating $\frac{dy}{dx}$ gives y.

You only need to look at the x bits, so you can put the numbers outside the integration signs.

Now just integrate both terms on the right-hand side.

Integrate each term separately...

$$y = \frac{1}{2}\int x^2\,dx - 3\int x^{-\frac{1}{2}}\,dx$$

The integration formula:

$$\int x^n\,dx = \frac{x^{n+1}}{n+1} + C$$

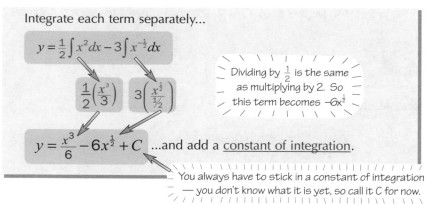

$$\frac{1}{2}\left(\frac{x^3}{3}\right) \qquad 3\left(\frac{x^{\frac{1}{2}}}{\frac{1}{2}}\right)$$

Dividing by $\frac{1}{2}$ is the same as multiplying by 2. So this term becomes $-6x^{\frac{1}{2}}$

$$y = \frac{x^3}{6} - 6x^{\frac{1}{2}} + C \quad \text{...and add a \underline{constant of integration}.}$$

You always have to stick in a constant of integration — you don't know what it is yet, so call it C for now.

You might wonder why there's only one constant. Since you've had to do three separate mini-integrations, shouldn't there be three constants like this?

$$y + C_1 = \frac{x^3}{6} + C_2 - 6x^{\frac{1}{2}} + C_3$$

Well yes, but rather than have three separate ones, you can just make them into one.

Let $C = -C_1 + C_2 + C_3 \Rightarrow y = \frac{x^3}{6} - 6x^{\frac{1}{2}} + C$

Paper 1 Q4 — Calculus

Now you've got to find C. The question says it's got to 'pass through the point $\left(1, \frac{1}{6}\right)$.'

<u>In other words</u>: when $x = 1$, $y = \frac{1}{6}$. Sticking this in will give you the value of C:

Substitute $x = 1$ and $y = \frac{1}{6}$ in the expression for y...

$$\frac{1}{6} = \frac{1^3}{6} - \left(6 \times 1^{\frac{1}{2}}\right) + C$$
$$= \frac{1}{6} - 6 + C$$
$$\Rightarrow C = 6$$

And so the complete expression for y is...

$$y = \frac{x^3}{6} - 6x^{\frac{1}{2}} + 6$$

You should check your answer by:

(i) putting in $x = 1$ (and making sure you get $\frac{1}{6}$)

and (ii) differentiating it (and making sure you get the derivative in the question).

(ii) A simple **Limit Integral** — but don't fall into the **Trap**

'Evaluate $\int_0^1 y\,dx$.'

This part is pretty standard stuff — but there's a really obvious trap at the start. As long as you avoid that, you can't go far wrong.

The Trap:
$$\int_0^1 y\,dx = \left[\frac{y^2}{2}\right]_0^1 = \frac{1}{2} - 0 = \frac{1}{2}$$

This is rubbish! — because this is integrating with respect to y, not x.

The Right Way:

$$\int_0^1 y\,dx.$$

The 'dx' means you're integrating <u>with respect to x</u>. So you need to write y <u>in terms of x</u> before you integrate. (i.e. stick in the answer to part (i).)

$$\int_0^1 y\,dx = \int_0^1 \left(\frac{x^3}{6} - 6x^{\frac{1}{2}} + 6\right)dx$$

You could break this up into individual chunks like before — but you don't have to. Do whatever's easier.

It's a limit integral, so integrate the bracket — and stick it in a big square bracket with limits.

$$\int_0^1 y\,dx = \int_0^1 \left(\frac{x^3}{6} - 6x^{\frac{1}{2}} + 6\right)dx = \left[\frac{x^4}{6 \times 4} - \frac{6x^{\frac{3}{2}}}{\frac{3}{2}} + \frac{6x^1}{1}\right]_0^1$$
$$= \left[\frac{x^4}{24} - 4x^{\frac{3}{2}} + 6x\right]_0^1$$

$$\frac{6}{\frac{3}{2}} = 6 \times \frac{2}{3} = \frac{12}{3} = 4$$

Now evaluate the square bracket — use the <u>top limit</u> first, then <u>subtract</u> what you get when you use the <u>bottom limit</u>.

$$\int_0^1 y\,dx = \left(\frac{1^4}{24} - \left(4 \times 1^{\frac{3}{2}}\right) + (6 \times 1)\right) - \left(\frac{0^4}{24} - \left(4 \times 0^{\frac{3}{2}}\right) + (6 \times 0)\right)$$
$$= \frac{1}{24} - 4 + 6$$
$$= \frac{49}{24}$$

When you put x = 0, all these parts are equal to zero.

$$\frac{1}{24} - 4 + 6 = \frac{1}{24} + 2 = \frac{48+1}{24} = \frac{49}{24}$$

Integration — Int it great? ...no? Oh... didn't think so...

This question is a gift — it's all real standard stuff. So if you're struggling with it, bury your head in those maths books until it begins to make sense. And another thing — always make sure you use all the info the question gives you, e.g. if it says the graph passes through the point (joe, bloggs), it means "at some point in the question, you need to plug in the values y=bloggs when x=joe". And watch for that trap at the end — the 'dx' in an integral means you have to integrate x's.

Paper 1 Q5 — Simultaneous Equations

> 5   **(i)**   Sketch the curve $y = (x-2)(x-4)$ and the line $y = 2x-4$ on the same set of axes, clearly
> marking the coordinates of the points of intersection.   **[3]**
>
>   **(ii)**   Evaluate the integral $\int_2^4 (x-2)(x-4)\,dx$.   **[3]**
>
>   **(iii)**   Hence or otherwise, show the total area enclosed by the lines $y = (x-2)(x-4)$ and $y = 2x-4$ is $\frac{32}{3}$.   **[4]**

(i)   Solve some *Simultaneous Equations* and *Sketch a Curve*

'Sketch the curve $y = (x-2)(x-4)$ and the line $y = 2x-4$ ...'

The question says you have to mark in the coordinates of the points where the parabola and the straight line cross. It's probably a good idea to find these <u>before</u> you draw anything — and that means solving simultaneous equations.

So solve these simultaneous equations to find the intersection points.

$$y = (x-2)(x-4) \quad \text{—} \quad ①$$
$$y = 2x-4 \quad \text{—} \quad ②$$

It's best to label them first.

Simultaneous equations where one of them is quadratic — straight away, think <u>substitution</u>. So...

Ⓐ Substitute y from equation 2 into equation 1:

$$2x-4 = (x-2)(x-4)$$
$$\Rightarrow 2x-4 = x^2 - 6x + 8$$

Rearrange things so that everything is on one side and you get...

$$x^2 - 8x + 12 = 0$$
$$\Rightarrow (x-2)(x-6) = 0$$
$$\Rightarrow x = 2 \quad or \quad x = 6$$

These are the x-coordinates of the points of intersection.

Now find the y-coordinates:

$$x = 2 \text{ in equation } 2 \Rightarrow y = (2 \times 2) - 4 = 0$$
$$x = 6 \text{ in equation } 2 \Rightarrow y = (2 \times 6) - 4 = 8$$

So the two points of intersection are: (2, 0) and (6, 8).

Ⓑ Then drawing the graph is easy. The parabola crosses the x-axis at x = 2 and x = 4, and it crosses the line at x = 2 and x = 6.

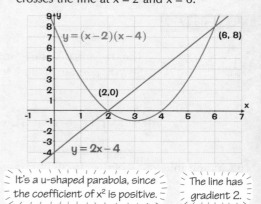

It's a u-shaped parabola, since the coefficient of x² is positive.   *The line has gradient 2.*

(ii)   A pretty easy *Integration*

'Evaluate the integral $\int_2^4 (x-2)(x-4)\,dx$.'

A pretty standard integration — shouldn't cause too many problems really...

$$\int_2^4 (x-2)(x-4)\,dx = \int_2^4 (x^2 - 6x + 8)\,dx$$
$$= \left[\frac{x^3}{3} - 3x^2 + 8x \right]_2^4$$
$$= \left(\frac{64}{3} - 48 + 32 \right) - \left(\frac{8}{3} - 12 + 16 \right)$$
$$= \frac{64}{3} - 48 + 32 - \frac{8}{3} + 12 - 16$$
$$= \frac{56}{3} - 20 = \frac{56-60}{3} = -\frac{4}{3}$$

Use the 'up the power by one and then divide by this new power' rule for each term.

Substituting in the limits and subtracting the results.

There's a minus sign in front of the brackets — so don't rush.

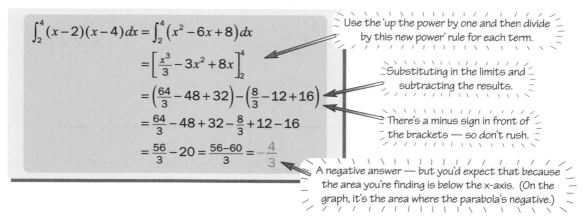

A negative answer — but you'd expect that because the area you're finding is below the x-axis. (On the graph, it's the area where the parabola's negative.)

Paper 1 Q5 — Simultaneous Equations

(iii) | Finding the **Area** between two **Curves** — use your **Sketch**

'Hence or otherwise, show the total area enclosed by the lines $y = (x-2)(x-4)$ and $y = 2x-4$ is $\frac{32}{3}$.'

This looks a bit tricky. Have another look at the graph to work out exactly what you've got to do.

You need the total area between the curves. This is: the green area B + the yellow area A.

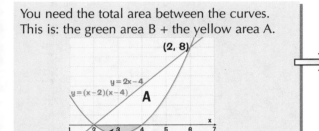

You've just found the green area B.
Now you need to work out the yellow area A.

The yellow area is:

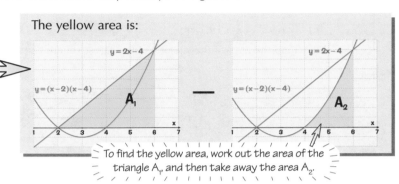

To find the yellow area, work out the area of the triangle A_1, and then take away the area A_2.

① The area of the triangle A_1 is given by:

$$A_1 = \tfrac{1}{2} \times base \times height$$
$$= \tfrac{1}{2} \times 4 \times 8 = 16$$

② The area you need to subtract, A_2, is:

$$A_2 = \int_4^6 (x-2)(x-4)\,dx = \int_4^6 (x^2 - 6x + 8)\,dx$$
$$= \left[\frac{x^3}{3} - 3x^2 + 8x\right]_4^6$$
$$= \left(\frac{216}{3} - 108 + 48\right) - \left(\frac{64}{3} - 48 + 32\right)$$
$$= \frac{152}{3} - 44 = \frac{152-132}{3} = \frac{20}{3}$$

③ So the yellow area A is:

$$A = A_1 - A_2$$
$$= 16 - \frac{20}{3} = \frac{48-20}{3} = \frac{28}{3}$$

④ And the total area enclosed by the two curves is:

$$Total\ area = yellow\ area + green\ area$$
$$= \frac{28}{3} + \frac{4}{3} = \frac{32}{3}$$

The integral you found in part (ii) had a minus sign — but you have to ignore it here. The minus sign tells you that the area is below the x-axis, but that's not important here — you just need to know the actual area.

⑨

But there's another way to find the **Area** between two **Curves**...

There is a quicker way to do this — but it's not as easy to see why it works.

If you 'subtract' the lower curve from the one that's higher up, and then integrate what you get — you find the area between them.

Between x = 2 and x = 6, the line is higher than the parabola, so take the equation of the parabola from the equation of the straight line.

$$Area\ between\ the\ curves = \int_2^6 \{(top\ curve) - (bottom\ curve)\}\,dx$$
$$= \int_2^6 \{(2x-4) - (x^2 - 6x + 8)\}\,dx$$
$$= \int_2^6 (-x^2 + 8x - 12)\,dx$$
$$= \left[-\frac{x^3}{3} + 4x^2 - 12x\right]_2^6$$
$$= \left(-\frac{216}{3} + 144 - 72\right) - \left(-\frac{8}{3} + 16 - 24\right)$$
$$= -\frac{208}{3} + 80 = \frac{-208+240}{3} = \frac{32}{3}$$

AS Maths: You shouldn't have done it if you can't take a joke...

There's nothing massively hard in this question, but there are bits that might catch you out if you're not paying attention, like that stuff about ignoring the minus sign (which is pretty confusing, I admit). The bit at the end about 'subtracting the curve from the line' is just a quick way to do the same stuff that's on the top half of the page (i.e. finding the area between two lines). If you don't get it, it doesn't matter — the top way's easier anyway. But it might save a bit of time...

Paper 1 Q6 — Sector Areas

6   The diagram below shows a sector of a circle of radius r cm and angle 120°.
The length of the arc of the sector is 40 cm.

   (i)   Write 120° in radians.   [1]

   (ii) Show that $r \approx 19.1$ cm.   [2]

   (iii) Find the area of the sector to the nearest square centimetre.   [3]

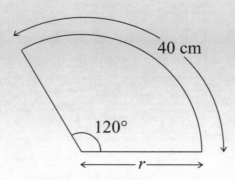

40 cm

120°

← r →

(i)   Use 360° = 2π Radians to Convert from Degrees to Radians

'Write 120° in radians.'

The first thing to notice is that the question says 'Write...'. This means you don't need to do any complicated working out — you can just give your answer in terms of π.

Now lots of people get in a tizz with **conversion factors**, but you will have to do them, so it's a good idea to have a **sure-fire method** up your sleeve. This is the way I do it:

1) You know that 360° = 2π, so the conversion factor is either going to be $\frac{2\pi}{360}$ or $\frac{360}{2\pi}$.

2) You're converting 120° into radians, so you're going to expect a **smaller** number than 120 when you've done the conversion.

3) If you want the conversion factor to give you a smaller number, it must be less than 1 (i.e. the top of the fraction must be smaller than the bottom). Well, $2\pi \approx 2 \times 3 = 6$, which is obviously less than 360, so

    you need to use the **first** conversion factor $\left(\frac{2\pi}{360}\right)$:

$$120° = 120° \times \frac{2\pi}{360°} \text{ radians}$$

$$= \frac{120°}{360°} \times 2\pi \text{ radians}$$

$$= \frac{1}{3} \times 2\pi \text{ radians}$$

$$= \frac{2\pi}{3} \text{ radians}$$

The formulas you're going to use for parts (ii) and (iii) need the angles to be measured in radians, so the first part of the question sets this up.

Paper 1 Q6 — Sector Areas

(ii) Remember the formula

'Show that $r \approx 19.1$ cm.'

You know the angle (θ) and the arc length (S), so you can use this formula to find the radius (r):

$$S = r \cdot \theta$$

Substituting S = 40 cm and $\theta = \dfrac{2\pi}{3}$ gives:

$$40 = \frac{2\pi}{3} \times r$$

Then rearrange to make r the subject:

$$r = 40 \div \frac{2\pi}{3}$$
$$= 40 \div 2.094$$
$$= 19.1 \text{ cm (3 s.f.)}$$

Don't throw marks away by forgetting the units.

(iii) Now use the other Formula to find the Area of the Sector

'Find the area of the sector to the nearest square centimetre.'

You've got the radius from part (ii), so you can stick it in the formula for the area of the sector...

...which, as I'm sure you've remembered, is this:

$$A = \frac{1}{2} r^2 \theta$$

The more decimal places you keep in your calculations, the more accurate your answer will be.

$r = 19.1$, $\theta = \dfrac{2\pi}{3}$ radians

$$A = \frac{1}{2} \times (19.099)^2 \times \frac{2\pi}{3}$$
$$= \frac{1}{2} \times 364.77 \times 2.094$$
$$= 382 \text{ cm}^2 \text{ (to the nearest cm}^2)$$

Remember, you still need to use the angle in radians.

If you can't remember which formula's which between '$r\theta$' and '$\frac{1}{2}r^2\theta$', think about which one's a **length** and which one's an **area**. The expression '$r\theta$' has just **one length** in it (r), so it must represent the arc **length** (S). The expression '$\frac{1}{2}r^2\theta$' has **two lengths** (r^2), so it must be the **area** of a sector.

Another handy tip...

If you're not sure if you've remembered the area formula properly, put in 2π for θ and make sure it gives you πr^2, because $2\pi = 360° = $ a full circle. You can also use this to work out the formula in reverse: $\dfrac{\theta}{2\pi} \times \pi r^2 = \dfrac{1}{2}r^2\theta$.

Alternatively, of course, you can just learn them, which might be easier.

Paper 1 Q7 — Algebra

7 **(i)** Rewrite the following expression in the form $f(x) = 0$, where $f(x)$ is of the form $f(x) = ax^3 + bx^2 + cx + d$.

$$(x-1)(x^2 + x + 1) = 2x^2 - 17$$

[2]

(ii) Show that $(x + 2)$ is a factor of $f(x)$. [3]

(iii) Using your answer to part (ii), factorise $f(x)$ as the product of a linear factor and a quadratic factor. [3]

(iv) By completing the square, or otherwise, show that $f(x) = 0$ has no other roots, and sketch the graph of $f(x)$. [2]

(v) Divide the polynomial $x^3 - 2x^2 + 3x - 3$ by $(x - 1)$, showing both the quotient and remainder. [4]

(i) *Multiply* out the *Brackets* and get everything on one side

'Rewrite the following expression in the form $f(x) = 0$...'

Looks confusing, but all it's asking you to do is multiply out the brackets and then rearrange it to get <u>zero</u> on one side.

$$(x-1)(x^2 + x + 1)$$
$$= x(x^2 + x + 1) - 1(x^2 + x + 1)$$
$$= x^3 + x^2 + x - x^2 - x - 1$$
$$= x^3 - 1$$

$$(x-1)(x^2 + x + 1) = 2x^2 - 17$$
$$\Rightarrow x^3 - 1 = 2x^2 - 17$$
$$\Rightarrow x^3 - 2x^2 + 16 = 0$$

Taking everything over to one side.

And this is in the form f(x) = 0, if f(x) is:

$$f(x) = x^3 - 2x^2 + 16$$

(ii) Show that something's a *Factor* — you need the *Factor Theorem*

'Show that $(x + 2)$ is a factor of $f(x)$.'

Whenever you see the word factor in a question — think '<u>Factor Theorem</u>'. There's ALWAYS a question on it. Which is good — cos it's easy.

To show whether (x + 2) is a factor of f(x), find f(–2)...

$$f(x) = x^3 - 2x^2 + 16$$
$$\Rightarrow f(-2) = (-2)^3 - 2 \times (-2)^2 + 16$$
$$= -8 - 8 + 16$$
$$= 0$$

Since f(–2) = 0, by the Factor Theorem, (x + 2) must be a factor of f(x).

THE FACTOR THEOREM

The Factor Theorem says that (x – a) is a factor of a polynomial f(x) if and only if f(a) = 0.

So if you want to show that (x+2) is a factor of f(x), just show that f(–2) = 0.

But don't get the plus and minus signs confused. To prove that (x+a) is a factor, show f(–a) = 0. To prove that (x–a) is a factor, show f(a) = 0.

(iii) Now you need to *Factorise a Cubic*

'Using your answer to part (ii), factorise $f(x)$ as the product of a linear factor and a quadratic factor.'

From part (ii), you know that (x + 2) is a factor of f(x). So...

Write down the factor you've already got, and put the x² term and the number in another set of brackets...

...then think about the term in the middle.

$$f(x) = x^3 - 2x^2 + 16$$
$$= (x+2)(\qquad)$$
$$= (x+2)(x^2 \qquad + 8)$$

These give you 2x²...

...so you need –4x² from these to give you the –2x² in f(x).

So put the –4x you need in the middle of the quadratic term and you get: $f(x) = (x+2)(x^2 - 4x + 8)$

Paper 1 Q7 — Algebra

(iv) *And for my next trick, I shall* Complete the Square...

'By completing the square, or otherwise, show that f(x) = 0 has no other roots, and sketch the graph of f(x).'
Now you've factorised f(x), the next bit's not too bad.

$$f(x) = (x+2)(x^2 - 4x + 8)$$
so $f(x) = 0$ when $(x+2) = 0$ or $(x^2 - 4x + 8) = 0$.

You've already shown that f(x)=0 when x + 2 = 0, so you need to check whether x² – 4x + 8 is ever zero.
The question gives you the hint that completing the square will be useful — so try that...

Completing the square with x² – 4x + 8:
$$x^2 - 4x + 8 = (x-2)^2 + something$$
$$= (x-2)^2 + 4$$

Since (x – 2)² can never be less than zero, the smallest value this can take is 4, and so it can definitely never be zero. And this means that f(x)=0 has no other solutions.

And now you know where f(x) is zero (only at x = –2), sketching the graph is easy.

Differentiate to find the turning points (if any).
$$f(x) = x^3 - 2x^2 + 16$$
$$\Rightarrow f'(x) = 3x^2 - 4x = x(3x - 4)$$
So the graph has turning points at x = 0 and x = $\frac{4}{3}$.

The coefficient of x³ is positive, so this is a bottom-left to top-right cubic.

The curve only crosses the x-axis once — at x = –2.

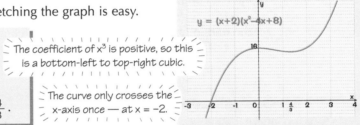
y = (x+2)(x²–4x+8)

(v) Algebraic Division — *Aaaaarrrrrrgggggghhhhhh....*

'Divide the polynomial x³ – 2x² + 3x – 3 by (x – 1)...'
Hmmm. Algebraic division. Don't fancy this much. Oh well, best get on with it.
You need to find $(x^3 - 2x^2 + 3x - 3) \div (x - 1)$ — so keep subtracting
lumps of (x – 1) to get rid of all the powers of x.

See page 1 for more about algebraic division.

First get rid of the x³ term by subtracting x² lots of (x – 1):
$$(x^3 - 2x^2 + 3x - 3) - x^2(x-1) = x^3 - 2x^2 + 3x - 3 - x^3 + x^2$$
$$= -x^2 + 3x - 3$$

Now do the same with the bit you've got left to get rid of the x² term. Add x lots of (x – 1), which is the same as subtracting –x lots of (x – 1):
$$(-x^2 + 3x - 3) + x(x-1) = (-x^2 + 3x - 3) - (-x)(x-1)$$
$$= -x^2 + 3x - 3 + x^2 - x$$
$$= 2x - 3$$

You're only interested in how many lumps of (x–1) you have to subtract, so rewrite it like this...

And finally, get rid of the x term in the bit that's left by subtracting 2 lots of (x – 1):
$$(2x - 3) - 2(x-1) = 2x - 3 - 2x + 2$$
$$= -1$$

And what that all means is this: $(x^3 - 2x^2 + 3x - 3) \div (x - 1) = x^2 - x + 2$ with remainder -1

This is the quotient...

...and this is the remainder.

The question's over — Am I not merciful... AM I NOT MERCIFUL...?

Now that was a question and a half. It was pretty long, but each of the individual bits was okay really. The main thing to take from these pages is that the Factor Theorem will definitely be in your exam. I kid you not — it will be there. And I reckon that anyone who knew that would definitely make sure they knew what the Factor Theorem was all about before the exam — especially since it's not even that hard. Think about it — guaranteed marks on a plate. Can't be bad.

Paper 1 Q8 — Circles

8   The diagram shows a circle. A (2, 1) and B (0, –5) lie on the circle and AB is a diameter.
C (4, –1) is also on the circle.

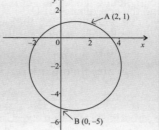

(i)   Find the centre and radius of the circle.   [4]

(ii) Show that the equation of the circle can be written in the form:

$$x^2 + y^2 - 2x + 4y - 5 = 0$$ [3]

(iii) The tangent at A and the normal at C cross at D.

Find the coordinates of D.   [7]

(i) | Average the Coordinates, then do a bit of Pythagoras

'Find the centre and radius of the circle.'

The centre of the circle must be at the midpoint of AB, since AB is a diameter.

> To get the midpoint, you **average the x- and y-coordinates**.
>
> Midpoint of AB is: $\left(\dfrac{2+0}{2}, \dfrac{1+-5}{2}\right) = (1, -2)$

As AB is the diameter, the length of the radius will be half the distance between A and B.

You can use Pythagoras' theorem to find the distance.
A quick sketch usually helps with these:

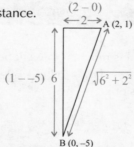

If you've forgotten about surds, see p 2.

$$AB = \sqrt{40} = 2\sqrt{10}$$
$$\text{Radius of the circle} = \frac{1}{2}AB = \sqrt{10}$$

(ii) | Use the General Equation of a Circle

'Show that the equation of the circle can be written in the form: $x^2 + y^2 - 2x + 4y - 5 = 0$'

The general equation for a circle with centre (a, b) and radius r is: $(x - a)^2 + (y - b)^2 = r^2$

> For this circle you have: $a = 1,\ b = -2,\ r = \sqrt{10}$
>
> $(x - 1)^2 + (y + 2)^2 = 10$

If we multiply out the brackets, you should be able to get the form given in the question:

> $(x - 1)(x - 1) + (y + 2)(y + 2) = 10$
> $x^2 - 2x + 1 + y^2 + 4y + 4 = 10$
> $x^2 + y^2 - 2x + 4y - 5 = 0$

Paper 1 Q8 — Circles

(iii) Tangents *touch* circles, Normals are at *right angles* to them

'Find the coordinates of D.'

A little bit of sketching on the diagram provided is always a good idea. ⇒

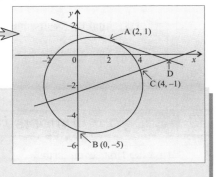

You need to find the equations of the tangent and normal and then work out where they cross.

The tangent at A:
This is at right angles to the radius and diameter at A — you can use this to find the gradient.

The gradient of a line joining (x_1, y_1) to (x_2, y_2) is given by:
$$\frac{y_2 - y_1}{x_2 - x_1}$$

For AB, gradient = $\frac{1 - -5}{2 - 0} = \frac{6}{2} = 3$

The **tangent is perpendicular to the diameter**, so use the gradient rule to find the gradient of the tangent.

$$\frac{-1}{\text{gradient of AB}} = -\frac{1}{3}$$

⟹

Use $y - y_1 = m(x - x_1)$ to get the equation of the tangent.

$m = -\frac{1}{3}$, (x_1, y_1) is the point (2, 1)

$$y - 1 = -\frac{1}{3}(x - 2)$$
$$y - 1 = -\frac{1}{3}x + \frac{2}{3}$$
$$y = -\frac{1}{3}x + \frac{5}{3}$$

So that's your equation for the tangent of the circle at point A.

The normal at C:
A normal passes through the centre, so its gradient is the gradient from the centre to C.

Gradient = $\frac{-1 - -2}{4 - 1} = \frac{1}{3}$

You can use: $y - y_1 = m(x - x_1)$ to get the equation of the normal.

Where $m = \frac{1}{3}$, (x_1, y_1) is the point (4, -1).

$$y - -1 = \frac{1}{3}(x - 4)$$
$$y + 1 = \frac{1}{3}x - \frac{4}{3}$$
$$y = \frac{1}{3}x - \frac{7}{3}$$

So here's your equation for the normal at point C.

Where they cross:
To get the coordinates of D, you need to find where the lines cross.

Put the equations equal to each other and solve:

$$-\frac{1}{3}x + \frac{5}{3} = \frac{1}{3}x - \frac{7}{3}$$
$$\frac{5}{3} + \frac{7}{3} = \frac{1}{3}x + \frac{1}{3}x$$
$$4 = \frac{2}{3}x$$
$$x = 6$$

Put this back into one of the equations to get y:

$$y = -\frac{1}{3}x + \frac{5}{3}$$
$$y = -\frac{1}{3} \times 6 + \frac{5}{3}$$
$$y = -2 + \frac{5}{3}$$
$$y = -\frac{1}{3}$$

So... D has coordinates $(6, -\frac{1}{3})$

Circles — a great shape for wheels...

OK, so at first glance this *looks* like a dirty great big question on circles. In fact, it's only half a dirty great big question on circles and half a dirty great big question on graphs. Tangents and normals questions are pretty much the same, whatever the shape of the curve you're looking at. All you need is the equation of the line and you're away. And if you notice, they give you the equation of the line in part (ii) — so you can still do part (iii), even if you make a pig's ear of the first two bits.

Paper 1 Q9 — Geometric Progressions

9 For the series with second term -2 and common ratio -½, find:

(i) the first seven terms [3]

(ii) the sum to infinity [3]

(i) Sub your numbers into the Formula for the nth term of a Geometric Series

'find the first seven terms'

The question gives you a couple of bits of information:

(i) The second term in the geometric series, $u_2 = -2$.

(ii) The common ratio, $r = -\dfrac{1}{2}$.

There are two ways of tackling this question.

Method 1: Work out the first term using the common ratio, then multiply the first term by the common ratio to get the second term, then multiply the second term by the common ratio to get the third term... etc.

OR **Method 2:** Work out the first term using the common ratio, then substitute that and the relevant n-value into the nth term formula — $ar^{(n-1)}$.

Find the first term...

The common ratio is $\dfrac{u_{n+1}}{u_n}$. It's the same for any pair of terms next to each other in the sequence — including the first and second terms.

So just plug in the values and solve for u_1:

$$\frac{u_2}{u_1} = \frac{u_2}{a} = -\frac{1}{2}$$

$$a = -2u_2 = -2(-2) = 4$$

The first term (u_1) is called 'a'

...then find the rest of them

With **Method 1**, each term is made up of the previous term multiplied by r.

With **Method 2**, each consecutive term is just the first term multiplied by a higher power of r

Term	n	Method 1	Method 2		Value
u_1	1	$a = 4$	$a = 4$	=	4
u_2	2	$ar = u_1 \times -\dfrac{1}{2} = 4 \times -\dfrac{1}{2}$	$ar = 4 \times \left(-\dfrac{1}{2}\right)^1$	=	-2
u_3	3	$(ar) \times r = u_2 \times -\dfrac{1}{2} = -2 \times -\dfrac{1}{2}$	$ar^2 = 4 \times \left(-\dfrac{1}{2}\right)^2$	=	1
u_4	4	$(ar^2) \times r = u_3 \times -\dfrac{1}{2} = 1 \times -\dfrac{1}{2}$	$ar^3 = 4 \times \left(-\dfrac{1}{2}\right)^3$	=	$-\dfrac{1}{2}$
u_5	5	$(ar^3) \times r = u_4 \times -\dfrac{1}{2} = -\dfrac{1}{2} \times -\dfrac{1}{2}$	$ar^4 = 4 \times \left(-\dfrac{1}{2}\right)^4$	=	$\dfrac{1}{4}$
u_6	6	$(ar^4) \times r = u_5 \times -\dfrac{1}{2} = \dfrac{1}{4} \times -\dfrac{1}{2}$	$ar^5 = 4 \times \left(-\dfrac{1}{2}\right)^5$	=	$-\dfrac{1}{8}$
u_7	7	$(ar^5) \times r = u_6 \times -\dfrac{1}{2} = -\dfrac{1}{8} \times -\dfrac{1}{2}$	$ar^6 = 4 \times \left(-\dfrac{1}{2}\right)^6$	=	$\dfrac{1}{16}$
u_n	n	$ar^{(n-2)} \times r$	$ar^{(n-1)}$	=	$ar^{(n-1)}$

1st term × r

2nd term × r

3rd term × r

Paper 1 Q9 — Geometric Progressions

(ii) | Simplify the numerator and plug in a and r

'find the sum to infinity'

Don't worry, this question sounds worse than it is. You <u>don't</u> have to work out every term from 1 to infinity (ha!) but you <u>will</u> have to use that formula for working out the sum to infinity of a geometric series (see p.24):

The tricky bit is remembering the formula:

> The sum to infinity of a converging geometric series is given by:
> $$S_\infty = \frac{a}{1-r}$$

After that, the rest of the question is just a case of plugging in the numbers and fiddling around again:

$a = 4$ and $r = -\frac{1}{2}$, so:

$$S_\infty = \frac{a}{1-r}$$

$$= \frac{4}{1-\left(-\frac{1}{2}\right)}$$

$$= \frac{4}{\frac{3}{2}} = \frac{2}{3} \times 4$$

$$= \frac{8}{3} = 2\frac{2}{3}$$

Check your answer looks right.

Look back at part (i) and work out the sums of the first 2, 3, 4, 5, 6 and 7 terms:

$S_1 = 4$

$S_2 = 4 - 2 = 2$

$S_3 = 4 - 2 + 1 = 3$

$S_4 = 4 - 2 + 1 - \frac{1}{2} = 2\frac{1}{2}$

$S_5 = 4 - 2 + 1 - \frac{1}{2} + \frac{1}{4} = 2\frac{3}{4}$

$S_6 = 4 - 2 + 1 - \frac{1}{2} + \frac{1}{4} - \frac{1}{8} = 2\frac{5}{8}$

$S_7 = 4 - 2 + 1 - \frac{1}{2} + \frac{1}{4} - \frac{1}{8} + \frac{1}{16} = 2\frac{11}{16}$

These seem to be settling down to somewhere between 2.5 and 2.75, so your answer looks feasible. Hurrah.

But if your mind goes blank in the exam and you can't remember the infinity sum equation, all's not lost. It's pretty easy to work out the formula for a sum to infinity from the general sum equation. All you need to do is look at what happens to the r^n, $\left(-\frac{1}{2}\right)^n$, term as n tends to ∞. The higher the n value, the more times you multiply $-\frac{1}{2}$ by itself. The more times you do this, the smaller the resulting number. $\left(-\frac{1}{2}\right)^\infty$ is a tiny, tiny number.

In fact, it's so small, you can just ignore it:

$$a(1 - 0) = a \implies S_\infty = \frac{a(1 - r_\infty)}{1 - r} = \frac{a}{1 - r}$$

...then you just carry on as before.

Sum to infinity — but not beyond...

OK, quick survey. Anyone actually *enjoy* this stuff? No offers? (sigh) Ah well. Can't say I disagree — if you ask me, questions like this are just a pain in the wotsits. But this area of maths does get more interesting if you decide to do a degree in it. And, of course, more *difficult*. Still — that's all part of the 'fun' of maths, isn't it. ☺

And it's no way, never. No way never, no more. And I'll do this geometric progression question... no never, no more.

General Certificate of Education
Advanced Subsidiary (AS) and Advanced Level

Core 2 Mathematics — Practice Exam Two

Give non-exact numerical answers correct to 3 significant figures, unless a different degree of accuracy is specified in the question or is clearly appropriate.

1 **(i)** Sketch the graph of $y = \cos(x - 60°)$ for x between $0°$ and $360°$. [2]

(ii) Show that the equation

$$2\sin^2(x - 60°) = 1 + \cos(x - 60°)$$ may be written as a quadratic in $\cos(x - 60°)$. [4]

(iii) Hence solve this equation, giving all values of x such that $0° \le x \le 360°$. [4]

2 **(i)** Write down the first four terms in the expansion of $(1 + ax)^{10}$, $a > 0$. [2]

(ii) Find the coefficient of x^2 in the expansion of $(2 + 3x)^5$. [2]

(iii) If the coefficients of x^2 in both expansions are equal, find the value of a. [2]

3 The diagram shows the graph of $y = 2^{x^2}$.

(i) Use the trapezium rule with 4 intervals to find an estimate for the area of the region bounded by the axes, the curve and the line x = 2. [3]

(ii) State whether the estimate in (i) is an overestimate or an underestimate, giving a reason for your answer. [2]

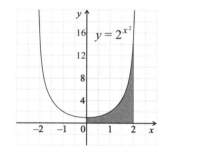

4 A geometric series $u_1 + u_2 + \dots + u_n$ has 3rd term $\frac{5}{2}$ and 6th term $\frac{5}{16}$.

(i) Find the common ratio and the first term of the series.

Hence give the formula for the nth term of the series. [4]

(ii) Find $\displaystyle\sum_{i=1}^{10} u_i$. Give your answer as a fraction in its simplest terms. [3]

(iii) Show that the sum to infinity of the series is 20. [2]

5   A new symmetrical mini-stage is to be built according to the design shown below. The design consists of a rectangle of length q metres and width $2r$ metres, two sectors of radius r and angle θ radians (shaded), and an isosceles triangle.

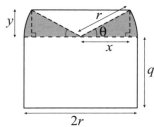

(i)   (a) Show that distance x is given by $x = r\cos\theta$ [1]

(b) Find a similar expression for distance y.   [1]

(ii) Find in terms of r, q and θ, expressions for the perimeter P, and the area A, of the stage.   [3]

(iii) If the perimeter of the stage is to be 40 metres, and $\theta = \frac{\pi}{3}$, show that A is given approximately by

$$A = 40r - 3.614r^2.$$ [4]

6 (i) Rewrite the following equation in the form $f(x) = 0$, where $f(x)$ is of the form $f(x) = ax^2 + bx + c$:

$$(x-1)(x-4) = 2x^2 + 11$$ [1]

(ii) By completing the square, or otherwise, show that $f(x) = 0$ has no real roots.   [2]

(iii) Sketch the graph of $f(x)$. Evaluate the area enclosed by the graph of $f(x)$, the line $y = \frac{1}{\sqrt{2}}$, the line $x = \sqrt{2}$ and the y-axis.   [4]

7   The circle with equation $x^2 - 6x + y^2 - 4y = 0$ crosses the y-axis at the origin and the point A.

(i)   Find the co-ordinates of A.   [2]

(ii) Rearrange the equation of the circle in the form: $(x-a)^2 + (y-b)^2 = c$.   [4]

(iii) Write down the radius and the co-ordinates of the centre of the circle.   [2]

(iv) Find the equation of the tangent to the circle at A.   [4]

8 (i) Sketch the graph of $y = 1 - (x-2)^2$, marking carefully the points where the curve meets the coordinate axes.   [3]

(ii) Using the same axes, sketch the graph of $y = 1.5x - 2$.   [2]

(iii) Prove that the x-coordinates of the points of intersection of the two graphs satisfy the equation $2x^2 - 5x + 2 = 0$. [2]

(iv) Solve this equation to find the coordinates of the points of intersection of the two graphs.   [3]

9 (i) Find the missing length a in the triangle.   [3]

(ii) Find the angles θ and ϕ.   [4]

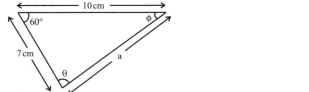

Paper 2 Q1 — Trigonometry

> **1 (i)** Sketch the graph of $y = \cos(x - 60°)$ for x between $0°$ and $360°$. [2]
>
> **(ii)** Show that the equation $2\sin^2(x - 60°) = 1 + \cos(x - 60°)$
> may be written as a quadratic in $\cos(x - 60°)$. [4]
>
> **(iii)** Hence solve this equation, giving all values of x such that $0° \le x \le 360°$. [4]

(i) | A **cos** graph shifted 60° to the **Right**

'Sketch the graph of $y = \cos(x - 60°)$ for x between $0°$ and $360°$.'

It's pretty easy to remember that you have to shift the graph <u>sideways</u> when the '−60°' is <u>inside</u> the brackets. But what's not so easy is to remember which <u>direction</u> to shift it — to the <u>left</u> or to the <u>right</u>.

It'll help you decide which way to move the graph if you remember this:

> $\cos x = 1$ when $x = 0$.
> So $\cos(x - 60°) = 1$ when $x - 60° = 0$ — and this is when <u>$x = 60°$</u>.

The graph is shifted horizontally by 60°. Because it's minus 60°, it's shifted to the right.

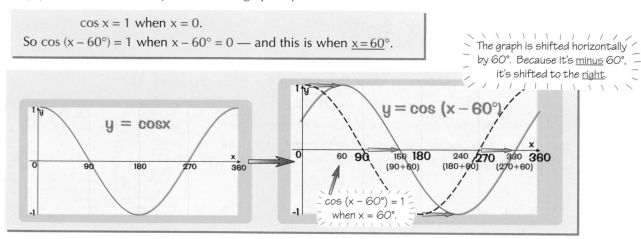

$\cos(x - 60°) = 1$ when $x = 60°$.

(ii) | Use **sin² + cos² = 1** to get rid of the **sin²**

'Show that the equation $2\sin^2(x - 60°) = 1 + \cos(x - 60°)$ may be written as a quadratic in $\cos(x - 60°)$.'

'A quadratic in $\cos(x - 60°)$...' — sounds a bit hard. But all you have to do is treat $\cos(x - 60°)$ like a single variable.

$$2\sin^2(x - 60°) = 1 + \cos(x - 60°) \quad —— ①$$

But '$\sin^2$(anything)$+\cos^2$(anything) = 1'. With a little rearranging, this means you can <u>replace</u> the $\sin^2$ with a $1 - \cos^2$.

You know from above that

$$2\sin^2(x - 60°) = 1 + \cos(x - 60°)$$

$\sin^2(x-60°) + \cos^2(x-60°) = 1$
$\Rightarrow \sin^2(x-60°) = 1 - \cos^2(x-60°)$

Now you can substitute for $\sin^2(x - 60°)$ in this to give...

$$2\{1 - \cos^2(x - 60°)\} = 1 + \cos(x - 60°)$$

You could even substitute y for x−60° — and work with sin²y and cos²y instead.

Now just multiply out the bracket and <u>rearrange</u> this so that everything's on one side.

$$2\{1 - \cos^2(x - 60°)\} = 1 + \cos(x - 60°)$$
$$\Rightarrow 2 - 2\cos^2(x - 60°) = 1 + \cos(x - 60°)$$
$$\Rightarrow 2\cos^2(x - 60°) + \cos(x - 60°) - 1 = 0$$

If you write y = cos(x − 60°), then this is 2y² + y − 1 = 0. That's what the question means by 'a quadratic in cos(x − 60°)'.

Paper 2 Q1 — Trigonometry

(iii) | It's a *Quadratic*, so try to *Factorise* it

'Hence solve this equation, giving all values of x such that $0° \leq x \leq 360°$.'

It's another question that looks a lot more frightening than it actually is. The most important thing is that you do it one bit at a time. You've got a quadratic equation (even though it's a horrible-looking one), so try to factorise it.

But it's probably a good idea to <u>rewrite</u> it a bit first so that it looks friendlier...

$$2\cos^2(x-60°) + \cos(x-60°) - 1 = 0$$

If you substitute y for cos(x – 60°), this becomes...

$$2y^2 + y - 1 = 0$$

Yep, it's a normal quadratic, so try to factorise it — and if it won't factorise, use the quadratic formula.

This quadratic factorises to give...

$$(2y-1)(y+1) = 0$$
$$\Rightarrow y = \tfrac{1}{2} \text{ or } y = -1$$

So you've found what y is. But y = cos(x – 60°), and so...

$$\cos(x-60°) = \tfrac{1}{2} \text{ or } \cos(x-60°) = -1$$

At this stage, alarm bells should definitely be ringing — you've just drawn the graph of y = cos(x – 60°), and you can use that to help solve this part.

And by taking the inverse cosine of these, you get...

$$x - 60° = \cos^{-1}\left(\tfrac{1}{2}\right) = 60°$$
$$\Rightarrow x = 120°$$

or

$$x - 60° = \cos^{-1}(-1) = 180°$$
$$\Rightarrow x = 240°$$

Get these values from your calculator.

So far, you've got two solutions — but there might be <u>more</u>. It's time to have another look at the graph from part 1. That's basically why they asked you to draw it — to <u>help</u> you with this part.

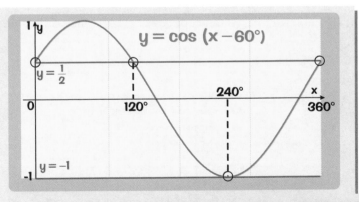

$$y = \cos(x - 60°)$$

From equation 2, you know that you're looking for points where either:

$$\cos(x-60°) = \tfrac{1}{2} \text{ or } \cos(x-60°) = -1.$$

Looking at the graph, there are <u>four</u> possible solutions — and you've already got two of them.

The other two solutions are at the extreme left and the extreme right, x = 0° and x = 360°.

So the four solutions are: x = 0°, x = 120°, x = 240° and x = 360°.

All these solutions are okay since you have to find solutions with $0° \leq x \leq 360°$.

Get a tan — cos you're worth it...

What a question — what a pain. It relies on knowing all sorts of things: graphs, $\sin^2 + \cos^2 = 1$, solving quadratics, solving trig equations, and more besides. The thing to remember, though, is that with questions like this, later parts often make use of stuff you worked out in the earlier sections — so if something seems outrageously difficult, have a look at what you've already done and see if you can get a hint from that.

Paper 2 Q2 — Binomial Expansion

2 **(i)** Write down the first four terms in the expansion of $(1 + ax)^{10}$, $a > 0$. [2]

(ii) Find the coefficient of x^2 in the expansion of $(2 + 3x)^5$. [2]

(iii) If the coefficients of x^2 in both expansions are equal, find the value of a. [2]

(i) | *Binomial Expansion — Use the Formula but watch out for the 'a'*

'Write down the first four terms in the expansion of $(1 + ax)^{10}$, $a > 0$.'

It's a <u>binomial expansion</u>.
That means you have two choices — Pascal's Triangle or the formula.

Now you *could* use Pascal's triangle — but it's a very <u>high power</u>, so it'd take up loads of valuable exam time when you could be worrying about another question.

So it's time to dig out the <u>formula</u>. (If you don't know it by now — learn it.)

There's more detail on the Binomial Expansion on page 24.

① Start by writing down the formula:

$$(1+x)^n = 1 + \frac{n}{1}x + \frac{n(n-1)}{1\times 2}x^2 + \frac{n(n-1)(n-2)}{1\times 2\times 3}x^3 + \ldots + \ldots + x^n$$

② Then write out the expression from the question...

$$(1 + ax)^{10}$$

Before you go any further, alarm bell should be ringing — you've got 'ax' instead of 'x'. Just remember that, OK.

③ ...expand it...

$$= 1 + \frac{10}{1}(ax) + \frac{10\times 9}{1\times 2}(ax)^2 + \frac{10\times 9\times 8}{1\times 2\times 3}(ax)^3 + \ldots$$

REMEMBER — square / cube the WHOLE BRACKET, not just the 'x'

④ ...and finally, simplify it.

$$= 1 + 10ax + \frac{90}{2}a^2x^2 + \frac{720}{6}a^3x^3 + \ldots$$

$$= 1 + 10ax + 45a^2x^2 + 120a^3x^3 + \ldots$$

So $(1 + ax)^{10}$ = 1 + 10ax + 45a²x² + 120a²x² + ...

Paper 2 Q2 — Binomial Expansion

(ii) Binomial Expansion again — Watch out for the '2'

'Find the coefficient of x^2 in the expansion of $(2 + 3x)^5$.'

(1) Again — start by writing down the formula:

$$(1+x)^n = 1 + \frac{n}{1}x + \frac{n(n-1)}{1 \times 2}x^2 + \frac{n(n-1)(n-2)}{1 \times 2 \times 3}x^3 + ... + ... + x^n$$

(2) Then write out the expression from the question.

$(2 + 3x)^5$ ⟵ *Again, look out. This time you've got a '2', not a '1', and you've got a '3x' instead of an 'x'.*

(3) You have to take a factor of 2 out to get it in the form you want:

$$(2+3x)^5 = \left[2\left(1+\tfrac{3}{2}x\right)\right]^5$$
$$= 2^5\left(1+\tfrac{3}{2}x\right)^5$$
$$= 32\left(1+\tfrac{3}{2}x\right)^5$$

Remember to take the 2 to the power 5, not just the bracket.

(4) Then expand it as before (only up to the x^2 term this time):

$$32\left(1+\tfrac{3}{2}x\right)^5 = 32\left[1 + \frac{5}{1}\left(\tfrac{3}{2}x\right) + \frac{5 \times 4}{1 \times 2}\left(\tfrac{3}{2}x\right)^2 + ...\right]$$

(5) You only need the x^2 term so just simplify that one:

Cancel as much as you can to make the calculation easier.

$$x^2 \text{ term} = 32 \times \frac{5 \times 4}{1 \times 2}\left(\tfrac{3}{2}x\right)^2$$

$$= \frac{\overset{16}{\cancel{32}} \times \overset{5}{\cancel{20}}}{1 \times \cancel{2}} \times \frac{9}{\cancel{4}_{1}}x^2 = 16 \times 5 \times 9x^2 = 80 \times 9x^2 = 720x^2$$

So the coefficient of x^2 is 720.

(iii) Equate coefficients from the first two answers — no problem

'If the coefficients of x^2 in both expansions are equal, find the value of a.'

If I was less of a cynic, I'd think they put in questions like this to be nice to you. But it's more likely they just put them in when they need a 2-mark question to make up the total marks. But whatever the reason, here's what you do:

The x^2 term from **(i)** was $45a^2x^2$ and the x^2 term from **(ii)** was $720x^2$

You're told that the coefficients are equal, so guess what you do...

...yep, you put them equal to each other! (genius) ⟹ $45a^2 = 720$

Then just fiddle around with it to solve for a: ⟹ $a^2 = 720 \div 45 = 16$

so $a = \pm 4$

But since it says in Question **(i)** that $a > 0$, you can confidently say that $a = 4$.

---

binomial expansion: $(1 + \text{binomial})^7 = 1 + 7\,\text{binomial} + 21\,\text{binomial}^2 + 35\,\text{binomial}^3 + ...$ *ho ho*

The problem with the binomial expansion is that it's a bit fiddly — there are loads of bits to it, and that means loads of opportunities for cocking it up. So the best advice I can give is to firstly *LEARN* the expansion, and secondly to take *EXTRA SPECIAL CARE* when using it. Check every step carefully until you're confident you haven't missed anything.

Paper 2 Q3 — Trapezium Rule

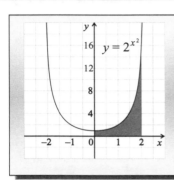

3 The diagram shows the graph of $y = 2^{x^2}$.

(i) Use the trapezium rule with 4 intervals to find an estimate for the area of the region bounded by the axes, the curve and the line $x = 2$. [3]

(ii) State whether the estimate in (i) is an overestimate or an underestimate, giving a reason for your answer. [2]

(i) **Make a Rough Sketch of the Four Trapezia**

'Use the trapezium rule with 4 intervals to find an estimate for the area of the region bounded by the axes, the curve and the line $x = 2$.'

Drawing even a very rough sketch of the graph can help you see what you're actually doing.
Copy the graph and add on the 4 strips.
Here's one I made earlier:

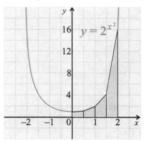

Look up the formula for the trapezium rule in the formula booklet and write it down:

$$\int_a^b y\,dx \approx \frac{h}{2}\left[y_0 + 2(y_1 + y_2 + \ldots + y_{n-1}) + y_n\right]$$

n is the number of intervals
h is the width of each strip

The width of each strip is 0.5 — you can get that from your sketch or by working out $h = \frac{2-0}{4} = 0.5$.

Draw up a table to work out the y-values.

x	$y = 2^{x^2}$
$x_0 = 0$	$y_0 = 2^{0^2} = 2^0 = 1$
$x_1 = 0.5$	$y_1 = 2^{0.5^2} = 2^{0.25} = 1.189$ (3 d.p.)
$x_2 = 1$	$y_2 = 2^{1^2} = 2^1 = 2$
$x_3 = 1.5$	$y_3 = 2^{1.5^2} = 2^{2.25} = 4.757$ (3 d.p.)
$x_4 = 2$	$y_4 = 2^{2^2} = 2^4 = 16$

Putting all these values into the formula gives:

$$\int_0^2 2^{x^2}\,dx \approx \frac{0.5}{2}\left[1 + 2(1.189 + 2 + 4.757) + 16\right]$$

$$= \frac{1}{4}(1 + 15.892 + 16)$$

$$= \frac{1}{4}(32.892)$$

$$= 8.22 \text{ (3 s.f.)}$$

As long as you're careful with all that working out, I'm sure you'll have no trouble.

Paper 2 Q3 — Trapezium Rule

(ii) *The shape of the curve tells you whether it's an overestimate or underestimate*

'State whether the estimate in (i) is an overestimate or an underestimate, giving a reason for your answer.'

Looking back at the sketch of the curve:

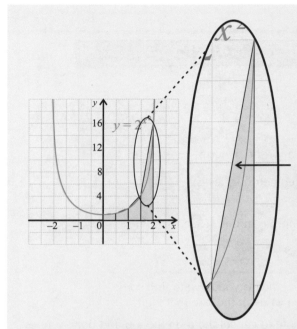

Each trapezium goes higher than the curve, so within each interval, the area of each trapezium is greater than the area under the curve.

The trapezium rule therefore gives an **overestimate** for the area.

Logarithm 'n' blues

Hate to do this to you, but I'm going to sneakily give you an extra question for practice... and it's on logs. I know, I know, I'm a wicked person, but I'm sure you'll rise to the challenge. It'll be over soon.

You can see on the graph of $y = 2^{x^2}$ that the line goes through the point (2, 16).

See if you can work backwards from $2^{x^2} = 16$, using logs, to show that $x = 2$ is a solution.

I'll give you a few seconds to think about it...

La la la, la la la-la la la la la, la la la-la la...   That's Kylie, by the way.

Right then.
Take logs of both sides (to the base 2 so you can bring that x^2 down):

Use the law of logs:

$2^{x^2} = 16$

$\log_2 2^{x^2} = \log_2 16$

$x^2 \log_2 2 = \log_2 16$ Spot that $\log_2 2$ is 1 (because $2^1 = 2$).

$x^2 = \log_2 16$ Spot that $\log_2 16$ is 4 (because $2^4 = 16$).

$x^2 = 4$

$x = \pm 2$ Remember the '±' — the graph is symmetrical about the y-axis.

All over. See how clever you are? You can take anything I throw at you.

Paper 2 Q4 — Geometric Series

4 A geometric series $u_1 + u_2 + \ldots + u_n$ has 3rd term $\frac{5}{2}$ and 6th term $\frac{5}{16}$.

(i) Find the common ratio and the first term of the series. Hence give the formula for the nth term of the series. **[4]**

(ii) Find $\displaystyle\sum_{i=1}^{10} u_i$. Give your answer as a fraction in its simplest terms. **[3]**

(iii) Show that the sum to infinity of the series is 20. **[2]**

(i) Put the numbers they give you into the Formula for the nth Term

'Find the common ratio and the first term of the series.'

So far, you've only got a couple of bits of information. You know:

        (i) that this is a <u>geometric</u> series,

and   (ii) two of the terms.

So you might as well start by putting these values into the formula for the nth term.

> The formula for the nth term of a geometric series is:
> $$u_n = ar^{n-1}$$
> And substituting the values for u_3 and u_6 in this gives:
> $$u_3 = ar^2 = \frac{5}{2} \qquad u_6 = ar^5 = \frac{5}{16}$$

You need to find values for a and r from these two equations. That means you have to find some way of getting rid of either a or r — then you'll be able to find the value of the one that's left.

The trick here is to <u>divide</u> these two expressions, then you'll be able to <u>cancel</u> the a on the top and bottom lines.

Divide the expressions for u_3 and u_6...

Cancel the a, and also an r^2, to leave just r^3. → $\dfrac{ar^5}{ar^2} = r^3 = \dfrac{5/16}{5/2}$ ← *An equation with just one unknown in — r.*

Now that you've got an equation just containing r, you can solve it.

$$r^3 = \frac{5/16}{5/2} = \frac{5}{16} \times \frac{2}{5} = \frac{\cancel{5}^1}{\cancel{16}^8} \cdot \frac{\cancel{2}^1}{\cancel{5}^1} = \frac{1}{8}$$

Cancel anything you can to make things easier.

So take the cube root, and you find...

$$r = \sqrt[3]{\frac{1}{8}} = \frac{1}{\sqrt[3]{8}} = \frac{1}{2}$$

See pages 21 - 23 for more about geometric series.

Substitute this value back into the equation for u_3 to find a...

$$ar^2 = \frac{5}{2}$$

You could use the equation for u_6, but this'll be easier because the power of r is smaller.

$$\Rightarrow a\left(\frac{1}{2}\right)^2 = \frac{1}{4}a = \frac{5}{2}$$

$$\Rightarrow a = 10$$

The hard work's done, you've found a and r — but the question asks you for the <u>formula for the nth term of the series</u>.

The nth term is given by $u_n = ar^{n-1}$, where $a = 10$ and $r = \frac{1}{2}$.

So,

$$u_n = 10 \cdot \left(\frac{1}{2}\right)^{n-1} = 10 \cdot \frac{1}{2^{n-1}} = \frac{10}{2^{n-1}}$$

Write your answer as simply as you can.

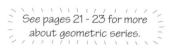

CHECK YOUR ANSWER:

You can check that your answer gives the numbers in the question.

Using $u_n = \dfrac{10}{2^{n-1}}$

$$u_3 = \frac{10}{2^2} = \frac{10}{4} = \frac{5}{2}$$

...so this one's okay. And

$$u_6 = \frac{10}{2^5} = \frac{10}{32} = \frac{5}{16}$$

...so this is fine too.

Paper 2 Q4 — Geometric Series

(ii) Now you need the Sum of the first Ten Terms

'Find $\sum_{i=1}^{10} u_i$.'

The Σ-notation just means that you're looking for a <u>sum</u> — so use the formula for the sum of the first n terms...

The sum of the first n terms of a <u>geometric series</u> is given by:

$$\sum_{i=1}^{n} u_i = S_n = \frac{a(1-r^n)}{1-r}$$

So put n = 10 to find the sum of the first ten terms, S_{10}...

$$S_{10} = \sum_{i=1}^{10} u_i = \frac{a(1-r^{10})}{1-r}$$

Then plug in the values you found for a and r in part (i)...

Using the values a = 10 and r = $\frac{1}{2}$ ...

$$S_{10} = 10 \times \frac{1-(1/2)^{10}}{1-1/2} = 10 \times \frac{1-1/2^{10}}{1/2} = 10 \times 2 \times \left(1 - 1/2^{10}\right)$$

Dividing by $\frac{1}{2}$ is the same as multiplying by 2.

You can tell if a fraction can be cancelled down further by looking at the <u>prime factors</u> of the top and bottom lines.

E.g. $\frac{5115}{256} = \frac{5115}{2^8}$

1) The bottom line is 2^8, so the <u>only</u> factors you could possibly cancel are 2, 4, 8, 16,... and so on.

2) As 2 doesn't go into 5115, neither do 4, 8, 16,... etc. — so there are no common factors to cancel down.

Now simplify this as much as you can...

$$S_{10} = 20 \times \left(1 - \frac{1}{1024}\right) = 20 \times \left(\frac{1023}{1024}\right) = 5 \times \left(\frac{1023}{256}\right)$$

$$= \frac{5115}{256}$$

$2^{10} = 1024$

Cancelling a factor of 4 in both 20 and 1024.

There are no more common factors, so this is your final answer.

(iii) And lastly, the Sum to Infinity

'Show that the sum to infinity of the series is 20.'

One bit to go, and they've told you what the answer should be — a comforting little check.

It's the sum to infinity this time — and that's easier than the previous bit.

The sum to infinity is given by...

$$S_\infty = \frac{a}{1-r}$$

This only works if –1 < r < 1.

So using the values a = 10 and r = $\frac{1}{2}$ ...

$$S_\infty = \frac{10}{1-1/2} = \frac{10}{1/2} = 10 \times 2 = 20$$

And that's handy, because it's the figure you're supposed to end up with.

IF YOU <u>DON'T</u> GET THE RIGHT ANSWER...
1) Check you haven't 'lost' any <u>minus signs</u> anywhere,
2) Make sure you've done any divisions involving fractions properly,
3) Check you've <u>copied</u> things down correctly from one line to the next.

If you really can't find your mistake, do another question and come back to this one <u>later</u> if you've got time.

Summertime... and the adding is easy...

There's really not that much they can ask you about geometric series — but they can dress the same stuff up in lots of different ways. Basically if you know three things: the formula for the nth term, the formula for the sum of the first n terms, and the formula for the sum to infinity, then you should be pretty well prepared to answer anything. One thing that's really important (and I mean REALLY important) is that you get used to deciding whether a series is arithmetic or geometric. For an <u>arithmetic</u> series, you <u>add</u> a number each time — and 'Arithmetic' and 'Add' both start with 'a'. For a <u>geometric</u> series, you <u>multiply</u> by a number each time — and 'Geometric' and 'Multiply' start with, er... 'g' and 'm'.

Paper 2 Q5 — Application of Calculus

5 A new symmetrical mini-stage is to be built according to the design shown below. The design consists of a rectangle of length q metres and width $2r$ metres, two sectors of radius r and angle θ radians (shaded), and an isosceles triangle.

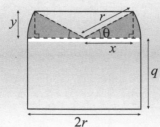

(i) (a) Show that distance x is given by $x = r\cos\theta$ [1]
    (b) Find a similar expression for distance y. [1]

(ii) Find in terms of r, q and θ, expressions for the perimeter P, and the area A, of the stage. [3]

(iii) If the perimeter of the stage is to be 40 metres, and $\theta = \frac{\pi}{3}$, show that A is given approximately by $A = 40r - 3.614r^2$. [4]

(i)   *Easy stuff with a **Right-Angled** triangle*

'Show that distance x is given by $x = r\cos\theta$ ...and find a similar expression for distance y.'

(a) It's easier if you draw a <u>picture</u> of the thing you're interested in. Here, it's the triangle in the top-right corner.

This is a <u>right-angled</u> triangle with an angle θ and hypotenuse r, and you need the length of the side <u>adjacent</u> to the angle — so use the cos formula...

$$\cos\theta = \frac{adjacent}{hypotenuse} = \frac{x}{r} \quad \text{...and so...} \quad x = r\cos\theta$$

(b) The question says that the stage is <u>symmetrical</u>, so although y is marked on the left-hand side of the picture, you can use your picture of the right-hand side. Distance y is the side <u>opposite</u> the angle — so use the sine formula...

$$\sin\theta = \frac{opposite}{hypotenuse} = \frac{y}{r} \quad \text{...and so...} \quad y = r\sin\theta$$

(ii)   *Find the Areas of **Sectors** and lengths of **Arcs***

'Find in terms of r, q and θ, expressions for the perimeter P, and the area A, of the stage.'
Again, it helps if you draw a picture so you can get a better idea of what's going on.

Do the perimeter first: the red bits are easy — but you need the lengths S too.
Since the shaded areas are <u>sectors</u> of circles, use the formula for the length of an <u>arc</u>...

This distance is $2x (= 2r\cos\theta)$.

The length of one arc S is given by: $S = r\theta$

So the total perimeter P is: $P = q + 2r + q + r\theta + 2r\cos\theta + r\theta$

$$= 2q + 2r + 2r\theta + 2r\cos\theta$$
$$= 2\{q + r(1 + \theta + \cos\theta)\}$$

This is in terms of r, q and θ, like the question asks for.

Now do the same sort of thing for the area — break it down into easier lumps. The total area A is given by...

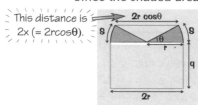

The area of the orange rectangle A_1:	The area of the orange triangle A_2:	The area of the grey sectors, each one having area A_3:
$A_1 = width \times height$	$A_2 = \frac{1}{2} \times width \times height$	Area of a sector: $A_3 = \frac{1}{2}r^2\theta$
$= 2r \times q$	$= \frac{1}{2}(2r\cos\theta)(r\sin\theta)$	
$= 2qr$	$= r^2\cos\theta\sin\theta$	$\Rightarrow$ Area of two sectors $= r^2\theta$

And so the final expression for the total area A is...

$$A = A_1 + A_2 + area\ of\ sectors$$
$$= 2qr + r^2\cos\theta\sin\theta + r^2\theta$$
$$= 2qr + r^2(\cos\theta\sin\theta + \theta)$$

Again, this is in terms of r, q and θ, like the question asks for.

See page 6-7 for more about lengths of arcs and areas of sectors.

Paper 2 Q5 — Application of Calculus

(iii) | *Put in the **Values** for P and θ — then **Fiddle About** a bit*

'...show that A is given approximately by $A = 40r - 3.614r^2$.'

Putting those values for P and θ into the formulas you've just found would be a good place to start...

Ⓐ Substitute for P and θ in the equation for the perimeter:

$\cos\frac{\pi}{3} = \frac{1}{2}$

$$P = 2\{q + r(1 + \theta + \cos\theta)\}$$

$$\Rightarrow 40 = 2\{q + r(1 + \frac{\pi}{3} + \cos\frac{\pi}{3})\}$$

$$\Rightarrow 20 = q + r(\frac{3}{2} + \frac{\pi}{3}) \quad —— ①$$

Ⓑ And then into the equation for the area:

$\cos\frac{\pi}{3} = \frac{1}{2}$

$\sin\frac{\pi}{3} = \frac{\sqrt{3}}{2}$

$$A = 2qr + r^2(\cos\theta\sin\theta + \theta)$$

$$= 2qr + r^2\left(\cos\frac{\pi}{3}\sin\frac{\pi}{3} + \frac{\pi}{3}\right)$$

$$= 2qr + r^2\left(\frac{1}{2}\cdot\frac{\sqrt{3}}{2} + \frac{\pi}{3}\right) = 2qr + r^2\left(\frac{\sqrt{3}}{4} + \frac{\pi}{3}\right) \quad —— ②$$

The formula for A in the question has an r and an r² on the right-hand side, and so does equation 2 — but you need to <u>get rid</u> of that q.

Ⓒ Rearranging equation 1 gives you...

$$q + r\left(\frac{3}{2} + \frac{\pi}{3}\right) = 20$$

$$\Rightarrow q = 20 - r\left(\frac{3}{2} + \frac{\pi}{3}\right)$$

Ⓓ Then use this to substitute for q in equation 2:

$$A = 2qr + r^2\left(\frac{\sqrt{3}}{4} + \frac{\pi}{3}\right)$$

$$= 2r\left\{20 - r\left(\frac{3}{2} + \frac{\pi}{3}\right)\right\} + r^2\left(\frac{\sqrt{3}}{4} + \frac{\pi}{3}\right)$$

$$= 40r - 2r^2\left(\frac{3}{2} + \frac{\pi}{3}\right) + r^2\left(\frac{\sqrt{3}}{4} + \frac{\pi}{3}\right)$$

$$= 40r - r^2\left\{2\left(\frac{3}{2} + \frac{\pi}{3}\right) - \left(\frac{\sqrt{3}}{4} + \frac{\pi}{3}\right)\right\}$$

This looks quite a lot like the thing you have to prove now...

When you take a negative factor outside the brackets (like −r²), all the signs <u>inside</u> the brackets change.

Ⓔ And since $2\left(\frac{3}{2} + \frac{\pi}{3}\right) - \left(\frac{\sqrt{3}}{4} + \frac{\pi}{3}\right) = 3.614$ to 3 d.p., this means: $\boxed{A = 40r - 3.614r^2}$

And one by one she had to tell them that her name was Joan of Arc...

What a question. It looks nasty when you first clap eyes on it, but as long as you keep your head and break it down into small, manageable chunks, it isn't really so bad. That's the way with a lot of these questions, but quite often the question tries to guide you through what you're supposed to do — so one big, difficult question becomes a few smaller, easier questionettes. The moral of this story is then: don't panic if the question looks impossible when you first see the exam paper — it's probably not so bad when you get down to it. And if it's still bad when you get down to it, use the Force...

Paper 2 Q6 — Graphs and Integration

6 **(i)** Rewrite the following equation in the form $f(x) = 0$, where $f(x)$ is of the form $f(x) = ax^2 + bx + c$:

$$(x-1)(x-4) = 2x^2 + 11$$ [1]

(ii) By completing the square, or otherwise, show that $f(x) = 0$ has no real roots. [2]

(iii) Sketch the graph of $f(x)$. Evaluate the area enclosed by the graph of $f(x)$, the line $y = \frac{1}{\sqrt{2}}$,

the line $x = \sqrt{2}$ and the y-axis. [4]

(i) The first part's Really Simple

'Rewrite the following equation in the form $f(x) = 0$, where $f(x)$ is of the form $f(x) = ax^2 + bx + c$:
$$(x-1)(x-4) = 2x^2 + 11$$'

The first thing to do is get rid of the brackets. So multiply them out:

$$x^2 - 5x + 4 = 2x^2 + 11$$

And rearrange, putting everything on one side (you want zero on the right-hand side).

$$-x^2 - 5x - 7 = 0$$
$$\Rightarrow x^2 + 5x + 7 = 0$$

And that's part (i) done. (So a=1, b=5, and c=7.)

(ii) Complete the Square and Use it

'By completing the square, or otherwise, show that $f(x) = 0$ has no real roots.'

So f(x) = x²+5x+7 from part (i). The <u>question</u> says to complete the square, so let's not argue.

Write the squared bracket down with d added.

$$\left(x + \tfrac{5}{2}\right)^2 + d$$ ← *d is some unknown number.*

The bracket is <u>always</u>:

$$a\left(x + \frac{b}{2a}\right)^2$$

You need to find a value for d such that $\left(x + \tfrac{5}{2}\right)^2 + d = f(x)$
So put in the old expression for f(x):

$$\left(x + \tfrac{5}{2}\right)^2 + d = x^2 + 5x + 7$$

<u>Always</u> put:
bracket ² + number = old expression for f(x).

and solve it to find d...

$$x^2 + 5x + \tfrac{25}{4} + d = x^2 + 5x + 7$$
$$d = 7 - \tfrac{25}{4} = \tfrac{3}{4}$$

The x² and x terms are the same on both sides so they cancel out. If they don't, you've got the bracket wrong.

So f(x) written in completed square form is:

$$f(x) = \left(x + \tfrac{5}{2}\right)^2 + \tfrac{3}{4}$$

You'll need to write something like this to show that you understand why it has no roots. Otherwise you won't get the marks.

You need to show that f(x)=0 has no roots.

$$f(x) = \left(x + \tfrac{5}{2}\right)^2 + \tfrac{3}{4} = 0$$

The squared bracket can <u>never</u> be less than 0. So the left-hand side can never be less than $\tfrac{3}{4}$. (So it can never be 0.) Therefore f(x) has <u>no roots</u>.

Paper 2 Q6 — Graphs and Integration

(iii) Sketching the Graph of y=f(x) — use the Completed Square

'Sketch the graph of $f(x)$.' $f(x) = x^2 + 5x + 7 \Rightarrow y = x^2 + 5x + 7$

To draw the graph you need to know where it has max/min points and where it crosses the axes.

The coefficient of x^2 is <u>positive</u> (it's actually 1), so it's going to be a <u>u-shaped</u> graph (rather than n-shaped).

Find where the min occurs (use the completed square form):

$$f(x) = y = \left(x + \tfrac{5}{2}\right)^2 + \tfrac{3}{4}$$

The minimum of f(x) is when the squared bracket is 0.

So 3/4 is the minimum value, and occurs when x=−5/2.

<u>Where does it cross the axes?</u>

It's <u>not</u> going to cross the x-axis, because its minimum value is ¾.
We can find where it crosses the y-axis by putting x=0:

$$x = 0 \Rightarrow y = 7$$

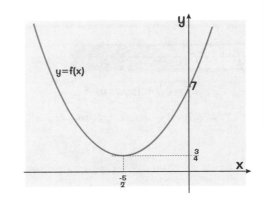

y=f(x)

Evaluate the Area — Use Your Sketch

'Evaluate the area enclosed by the graph of $f(x)$, the line $y = \frac{1}{\sqrt{2}}$, the line $x = \sqrt{2}$ and the y-axis.'

The first thing to do is to sketch the situation so you can see <u>exactly</u> what you've got to do.

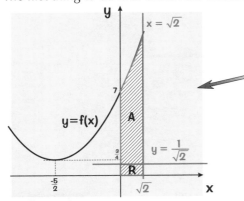

And draw on the 2 lines $x = \sqrt{2}$ and $y = \frac{1}{\sqrt{2}}$.

$\frac{1}{\sqrt{2}}$ is about 0.71, so it's just below $\frac{3}{4}$.

I've labelled two regions on the graph (A and R). Area A is the region you need to find. R is the small rectangular region underneath it.

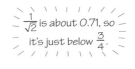

area A = area A+R − area R

Find this easily.

This is what we want.

Integrating y between 0 and $\sqrt{2}$ will give you the area A+R. (You find areas under curves by integrating. See pages 31.)

Find Area A+R:

$$\text{Area A+R} = \int_0^{\sqrt{2}} y\,dx = \int_0^{\sqrt{2}} \left(x^2 + 5x + 7\right)dx$$

$$= \left[\frac{x^3}{3} + \frac{5x^2}{2} + 7x\right]_0^{\sqrt{2}}$$

$$= \left(\frac{\sqrt{2}^3}{3} + \frac{5\sqrt{2}^2}{2} + 7\sqrt{2}\right) - \left(\frac{0^3}{3} + \frac{5\times 0^2}{2} + 7\times 0\right)$$

$$= \frac{2\sqrt{2}}{3} + \frac{5\times 2}{2} + 7\sqrt{2} \quad - \quad 0$$

$$= \sqrt{2}\left(\frac{2+21}{3}\right) + 5 = \frac{23}{3}\sqrt{2} + 5$$

Find Area R:

$$\text{Area R} = \text{length} \times \text{height} = \left(\sqrt{2} - 0\right) \times \left(\frac{1}{\sqrt{2}} - 0\right)$$

$$= \frac{\sqrt{2}}{\sqrt{2}} = 1$$

Find Area A:

$$\text{Area A} = \text{A+R} - \text{R}$$

$$= \frac{23}{3}\sqrt{2} + 5 - 1 \quad = \quad \frac{23\sqrt{2} + 4}{3}$$

Integrating graphs is an area that needs a lot of attention...

This is a pretty easy question, except for the last bit — and that's not <u>too</u> bad, once you've done the sketch. But if you've any sense, you'll check your answer... won't you. Your best bet with odd areas is to approximate it to easy shapes and check it's <u>about right</u>. For this question, it'd be a rectangle $\sqrt{2}$ by $\left(7 - \frac{1}{\sqrt{2}}\right)$ and a triangle $\sqrt{2}$ wide and roughly (16−7) tall.

Paper 2 Q7 — Circles

> 7   The circle with equation $x^2 - 6x + y^2 - 4y = 0$ crosses the y-axis at the origin and at the point A.
>
>   **(i)**   Find the coordinates of A.   [2]
>
>   **(ii)**   Rearrange the equation of the circle into the form: $(x - a)^2 + (y - b)^2 = c$.   [4]
>
>   **(iii)** Write down the radius and the coordinates of the centre of the circle.   [2]
>
>   **(iv)** Find the equation of the tangent to the circle at A.   [4]

(i) An Easy Start

'Find the coordinates of A.'

A is on the y-axis, so the x-coordinate is 0. Just put $x = 0$ into the equation and solve for y.

Here's the equation the question gives you:
$$x^2 - 6x + y^2 - 4y = 0$$

Putting in $x = 0$:
$$0^2 - 6 \times 0 + y^2 - 4y = 0$$
$$y^2 - 4y = 0$$

This will factorise:
$$y(y - 4) = 0$$
$$y = 0 \text{ or } 4$$

$y = 0$ is the origin, so at A, $y = 4$.   A is at $(0, 4)$.

(ii) This is like Completing the Square

'Rearrange the equation of the circle into the form: $(x - a)^2 + (y - b)^2 = c$.'

The first thing to do is to find the two terms in the brackets — a and b.

You're basically completing the square for x and y separately, so you have to **halve the number in front of x to get a, and halve the number in front of y to get b.**

But for each one you'll end up with a number that you don't want (— namely a^2 and b^2). You'll need to take it away each time.

So here's the equation you're starting from: $x^2 - 6x + y^2 - 4y = 0$

First deal with x: Half of −6 is −3. $x^2 - 6x = (x - 3)^2 - 9$ $(-3)^2 = 9$, so you need to take 9 away.

Now the same for y: Half of −4 is −2. $y^2 - 4y = (y - 2)^2 - 4$ $(-2)^2 = 4$, so you need to take 4 away.

You can now put these new expressions back into the original equation.

$$(x - 3)^2 - 9 + (y - 2)^2 - 4 = 0$$
$$(x - 3)^2 + (y - 2)^2 - 13 = 0$$
$$\boxed{(x - 3)^2 + (y - 2)^2 = 13}$$

Paper 2 Q7 — Circles

(iii) | What do *a*, *b* and *r* stand for?

'Write down the radius and the coordinates of the centre of the circle.'

When a question tells you to write something down, it means exactly that — no working's needed.
You do need to know some facts about the equation of a circle though.

> In the general equation for a circle $(x - a)^2 + (y - b)^2 = r^2$,
> the **centre** is (a, b) and the **radius** is *r*.

The only difference between this general equation of a circle and the form that you found in part (ii) is the r^2 instead of *c*. *c* must therefore equal r^2, so the radius (*r*) must be $\sqrt{c}$.

$a = 3$, $b = 2$, $r = \sqrt{c} = \sqrt{13}$

So the centre is $(3, 2)$, and the radius is $\sqrt{13}$.

(iv) | A Tangent just touches the Circle

'Find the equation of the tangent to the circle at A.'

The tangent is at right angles to the radius. This is one of those mega-important facts about circles that you've just gotta learn. You can find the gradient of the radius, and use that to find the gradient of the tangent.

The radius at A has a gradient of: $\dfrac{y_2 - y_1}{x_2 - x_1} = \dfrac{2-4}{3-0} = -\dfrac{2}{3}$

Use the points A $(0, 4)$ and C $(3, 2)$ as (x_1, y_1) and (x_2, y_2)

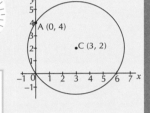

Remember the gradient rule? — Course you do.

> Gradient of line = $\dfrac{-1}{\text{gradient of perpendicular line}}$

So the tangent at A has a gradient of: $\dfrac{-1}{-\frac{2}{3}} = \dfrac{3}{2}$

To find the equation of the tangent you can use the old favourite: $y - y_1 = m(x - x_1)$

m is the gradient $\left(\dfrac{3}{2}\right)$, and (x_1, y_1) is the coordinate of a known point on the line — in this case A $(0, 4)$.

$$y - y_1 = m(x - x_1)$$
$$y - 4 = \frac{3}{2}(x - 0)$$
$$y - 4 = \frac{3}{2}x$$
$$y = \frac{3}{2}x + 4$$

Circles — very popular in late 60s fashion design...

You have to make sure you're totally OK with completing the square before you attempt part (ii) or you'll just get horribly confused. If you need to, go back to Section 2 and read up on completing the square, then come back to this one. This is a pretty tough question actually (soz like) so if you can do all this stuff with ease, you're pretty much on top of things.

Paper 2 Q8 — Graph Transformations

8 **(i)** Sketch the graph of $y = 1 - (x - 2)^2$, marking carefully the points where the curve meets the coordinate axes. [3]

   (ii) Using the same axes, sketch the graph of $y = 1.5x - 2$. [2]

   (iii) Prove that the x-coordinates of the points of intersection of the two graphs satisfy the equation

   $2x^2 - 5x + 2 = 0$. [2]

   (iv) Solve this equation to find the coordinates of the points of intersection of the two graphs. [3]

(i) Work out intercepts and turning point... then do a sketch

'Sketch the graph of $y = 1 - (x - 2)^2$, marking carefully the points where the curve meets the coordinate axes.'

Before you go any further, you want to <u>rearrange the expression</u> so that it's a clear '$ax^2 + bx + c$' quadratic.

> You can multiply through by –1 or leave the "–" sign outside the brackets so it's easier to factorise.

$$y = 1 - (x - 2)^2 = 1 - (x^2 - 4x + 4)$$
$$y = -(x^2 - 4x + 3)$$

To draw the graph you need to work out <u>three things</u>:

① the <u>x-intercepts</u>, where $y = 0$.

$0 = -(x^2 - 4x + 3)$
Factorising gives: $(x - 3)(x - 1) = 0$
So $x = 3$ or $x = 1$
So the graph cuts the x-axis at $x = 3$ and $x = 1$, i.e. at (3, 0) and (1, 0)

② the <u>y-intercept</u>, where $x = 0$.

$y = -[0^2 - (4 \times 0) + 3] = -3$
So the graph cuts the y-axis at $y = -3$, i.e. at (0, –3)

③ the <u>turning point</u>, where the <u>gradient = 0</u>.

Differentiate the expression for y to get the gradient:
$y = -(x^2 - 4x + 3) = -x^2 + 4x - 3$

$\dfrac{dy}{dx} = -2x + 4 = 0$ at turning point

which means $x = 2$ and $y = -[2^2 - (4 \times 2) + 3] = 1$

So, the turning point is at (2, 1).

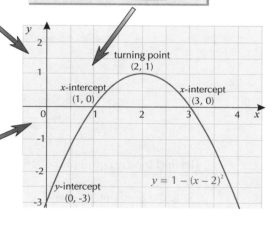

(ii) Sketch a **Straight** line — Easy

'Using the same axes, sketch the graph of $y = 1.5x - 2$.'

These are nice easy marks — all you have to do is <u>add a straight line to your graph</u>.

It's already in the form $y = mx + c$, so you know it's got a gradient of 1.5 and y-intercept of –2.

The easiest way to draw it is just plot two points and join them up. You've already got the y-intercept, (0, –2) so you need one more.

The gradient is 1.5 so count along 1 in the x-direction and up 1.5 — which takes you to (1, –0.5).

(Or you could do it twice so that you end up with both whole-number coordinates, (2, 1), but it's up to you.)

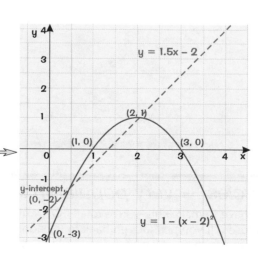

Paper 2 Q8 — Graph Transformations

(iii) Equate and rearrange

'Prove that the x-coordinates of the points of intersection of the two graphs satisfy the equation $2x^2 - 5x + 2 = 0$.'

When the x- and y-values of each graph are the same, it means that the two lines go through the <u>same point</u> — in other words they <u>cross</u>.

To find out where this happens, you treat them as <u>simultaneous equations</u>:

You've got 2 equations:
$$y = -(x^2 - 4x + 3) \quad — \text{(i)}$$
$$y = 1.5x - 2 \quad\quad — \text{(ii)}$$

At the point where they cross, the y-value of equation (i) is equal to the y-value of equation (ii), which gives:

Rearrange until you get what the question asks for:
$$-(x^2 - 4x + 3) = 1.5x - 2$$
$$-x^2 + 4x - 3 = 1.5x - 2$$
$$-x^2 + 2.5x - 1 = 0$$
$$2x^2 - 5x + 2 = 0 \quad \text{(exactly what you want)}$$

(iv) Solve a quadratic — wahey

'Solve this equation to find the coordinates of the points of intersection of the two graphs.'

On the home straight now — <u>solve the quadratic</u> in x (easy) then <u>substitute</u> the answer back in to find y. And I'm sure you're so familiar with solving quadratics by now that you could do them with your feet. But just in case...

Check whether it'll <u>factorise</u> (or just use the formula if you prefer).

This is the fiddly bit where you have to try a few combinations — check back in Core 1 if you've forgotten.

$$2x^2 - 5x + 2 = 0$$
$$(2x - 1)(x - 2) = 0$$
$$(2x - 1) = 0 \text{ or } (x - 2) = 0$$

So the two possible solutions are: $x = \dfrac{1}{2}$ and $x = 2$

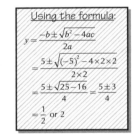
Using the formula:
$$y = \frac{-b \pm \sqrt{b^2 - 4ac}}{2a}$$
$$= \frac{5 \pm \sqrt{(-5)^2 - 4 \times 2 \times 2}}{2 \times 2}$$
$$= \frac{5 \pm \sqrt{25 - 16}}{4} = \frac{5 \pm 3}{4}$$
$$= \frac{1}{2} \text{ or } 2$$

Substitute these x-values back into the original equations to find the y-values.

The original equations were:
$$y = -(x^2 - 4x + 3) \quad — \text{(i)}$$
$$y = 1.5x - 2 \quad\quad — \text{(ii)}$$

For $x = \dfrac{1}{2}$

Equation (i) becomes:
$$y = -\left(\left(\frac{1}{2}\right)^2 - 4\left(\frac{1}{2}\right) + 3\right) = -\left(\frac{1}{4} - 2 + 3\right) = -\frac{5}{4}$$

Check with equation (ii):
$$y = 1.5\left(\frac{1}{2}\right) - 2 = \frac{3}{4} - 2 = -\frac{5}{4}$$

For $x = 2$

$$y = -(2^2 - (4 \times 2) + 3) = 1$$

$$y = 1.5(2) - 2 = 1$$

So, the points of intersection are $\left(\dfrac{1}{2}, -\dfrac{5}{4}\right)$ and $(2, 1)$

Look back at the intersection points on your graph to check they look about right.

What a beaut...

You see — this is the kind of question you really want to come up in your exam. The only part of that question that you couldn't do at GCSE was differentiating the quadratic to get the turning point. OK so you wouldn't get a question quite that long at GCSE, but most of the actual *maths* is <u>no harder</u>. So take comfort, O ye of little faith — AS Maths isn't *all* horrible.

Paper 2 Q9 — Sine and Cosine Rules

9   **(i)**   Find the missing length a in the triangle.   [3]

       (ii)   Find the angles θ and ϕ.   [4]

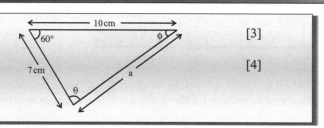

(i)   If you only know an **Angle** and the **Sides next to it**, use the **Cosine Rule**

'Find the missing length a...'

If you're asked to find a side or angle of a triangle without a right angle, there are two things you can try — the <u>Sine Rule</u> and the <u>Cosine Rule</u>. But to use the Sine Rule, you need to know both an angle and the length of the side opposite it. If you don't (like here), it'll have to be the Cosine Rule.

Use the Cosine Rule: $a^2 = b^2 + c^2 - 2bc \cos A$, putting b=10, c=7 and the angle A=60°.

$$a^2 = b^2 + c^2 - 2bc \cos A$$

Remember: Side a is opposite angle A, and so on.

$$a^2 = 10^2 + 7^2 - 2 \times 10 \times 7 \times \cos 60°$$

$$\Rightarrow a^2 = 100 + 49 - 140 \times 0.5$$

There's nothing really that tricky in the Cosine Rule — just loads of fiddly stuff all stuck together.

$$\Rightarrow a^2 = 149 - 70 = 79$$

$$\Rightarrow a = \sqrt{79} = 8.89 \, \text{cm}$$

Don't forget the units.

It doesn't matter which side you choose to be b or c. Just stick to whatever you *do* choose.

(ii)   Use the **Sine Rule** to find the missing angles

You've found side a in the first part of the question — so you know an angle and the length of the side opposite. Time to use the <u>Sine Rule</u> then. Don't worry — it's easier than falling off a dog.

'Find the angles θ...'

Now **θ** is the angle opposite the 10 cm side, and I called the 10 cm side b. So in the Sine Rule, **θ** is the angle B.

Put the values you know into the Sine Rule.

$$\frac{a}{\sin A} = \frac{b}{\sin B}$$

$$\frac{\sqrt{79}}{\sin 60°} = \frac{10}{\sin B}$$

Take sin B over to the left, and everything else to the right.

$$\Rightarrow \sin B = \frac{10 \times \sin 60°}{\sqrt{79}} = 0.974$$

$$\Rightarrow B = \sin^{-1} 0.9744 = 77.0°$$

Try to keep the full decimal answers on your calculator throughout the calculation. You should only actually round your answer at the very end.

The <u>Sine Rule</u>: In <u>any</u> triangle,
$$\frac{a}{\sin A} = \frac{b}{\sin B} = \frac{c}{\sin C}$$

The <u>Cosine Rule</u>: In <u>any</u> triangle,
$$a^2 = b^2 + c^2 - 2bc \cos A$$

'...and ϕ.'

Do exactly what you've just done. But use side c instead of b.

Put the values you know into the sine rule.

$$\frac{a}{\sin A} = \frac{c}{\sin C}$$

Rearrange things a bit.

$$\Rightarrow \sin C = \frac{7 \times \sin 60°}{\sqrt{79}} = 0.6820$$

And again — take the inverse sine to find the angle.

$$\Rightarrow C = \sin^{-1} 0.6820 = 43.0°$$

(Or you could just subtract the 2 angles you know from 180° — which is a good check.)

So θ = 77.0° and ϕ = 43.0°.

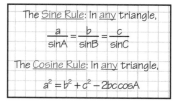

The Sine and Cosine Rules — I know what you're thinking...

The Sine and Cosine Rules are pretty darn easy, let's face it. If you get a question in the exam on them, you should be jumping for joy. They're both really flexible — for a triangle ABC, you could use any of these for the Cosine Rule: $a^2 = b^2 + c^2 - 2bc \cos A$, $b^2 = a^2 + c^2 - 2ac \cos B$, $c^2 = a^2 + b^2 - 2ab \cos C$. And for the Sine Rule, you can <u>either</u> put the lengths or the sine bits on the top: So either write it $\frac{a}{\sin A} = \frac{b}{\sin B} = \frac{c}{\sin C}$ or $\frac{\sin A}{a} = \frac{\sin B}{b} = \frac{\sin C}{c}$. Wow!

Answers

Section One — Algebra and Functions

1)a) $f(x) = (x + 2)(3x^2 – 10x +15) – 36$

 b) $f(x) = (x + 2)(x^2 – 3) +10$

2)a) (i) *You just need to find f(–1).*
 This is $–6 – 1 + 3 – 12 = –16$.
 (ii) *Now find f(1). This is* $6 – 1 – 3 – 12 = –10$.

 b) (i) $f(–1) = –1$
 (ii) $f(1) = 9$

3)a) *You need to find f(–2). This is*
 $(–2)^4 – 3(–2)^3 + 7(–2)^2 –12(–2) + 14$
 $= 16 + 24 + 28 + 24 + 14 = \underline{106}$

 b) *The remainder when you divide by (2x + 4) is the same as the remainder when you divide by x + 2, so the remainder is* $\underline{106}$.

4)a) *You need to find f(1) — if f(1) = 0, then (x – 1) is a factor:*
 $f(1) = 1 – 4 + 3 + 2 – 2 = 0$, so $\underline{(x – 1)\ is\ a\ factor}$.

 b) *You need to find f(–1) — if f(–1) = 0, then (x + 1) is a factor:*
 $f(–1) = –1 – 4 – 3 + 2 – 2 = –8$, so $\underline{(x + 1)\ is\ not\ a\ factor}$.

 c) *You need to find f(2) — if f(2) = 0, then (x – 2) is a factor:*
 $f(2) = 32 – (4 \times 16) + (3 \times 8) + (2 \times 4) – 2$
 $= 32 – 64 + 24 + 8 – 2 = –2$, so $\underline{(x – 2)\ is\ not\ a\ factor}$.

5) *If* $f(x) = 2x^4 + 3x^3 + 5x^2 + cx + d$, *then to make sure f(x) is exactly divisible by (x – 2)(x + 3), you have to make sure* $f(2) = f(–3) = 0$.
 $f(2) = 32 + 24 + 20 + 2c + d = 0$, *i.e.* $\underline{2c + d = –76}$.
 $f(–3) = 162 – 81 + 45 – 3c + d = 0$, *i.e.* $\underline{3c – d = 126}$.
 Add the two underlined equations to get: $5c = 50$, *and so* $\underline{c = 10}$. *Then* $\underline{d = –96}$.

Section Two — Circles and Trigonometry

£100   *See page 8*

£200   *See page 7*

£300   *See page 8*

£500   **a)** $B=125°$, $a=3.66$ m, $c=3.10$ m, *area is* 4.64 m^2

   b) $r=20.05$ km, $P=1.49°$, $Q=168.51°$

£1000   *Freda's angles are* $22.3°$, $49.5°$, $108.2°$

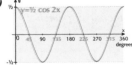

£2000   *One triangle:* $c=4.98$, $C=72.07°$, $B=72.93°$
   Other possible triangle: $c=3.22$, $C=37.93°$, $B=107.07°$

£4000   *See page 10*

£8000 a)

b)

c)

£16 000 **i)** **a)** $\theta = 240°, 300°$.
    b) $\theta = 135°, 315°$.
    c) $\theta = 135°, 225°$.

   ii) **a)** $\theta = 33.0°, 57.0°, 123.0°, 147.0°$,
       $-33.0°, -57.0°, -123.0°, -147.0°$
    b) $\theta = -17.5°, 127.5°$ **c)** $\theta = 179.8°$

£32 000   **a)** 3, (0, 0) **b)** 2, (2, –4) **c)** 5, (–3, 4)

£64 000 $x = 70.5°, 120°, 240°, 289.5°$.

£125 000 $x = -30°$

£250 000 $\left(\sin y + \cos y\right)^2 + \left(\cos y - \sin y\right)^2$
   $\equiv \left(\sin^2 y + 2\sin y \cos y + \cos^2 y\right) + \left(\cos^2 y - 2\cos y \sin y + \sin^2 y\right)$
   $\equiv 2(\sin^2 y + \cos^2 y) \equiv 2$

£500 000 $\dfrac{\sin^4 x + \sin^2 x \cos^2 x}{\cos^2 x - 1} \equiv -1$
   LHS:
   $\equiv \dfrac{\sin^2 x \left(\sin^2 x + \cos^2 x\right)}{\left(1 - \sin^2 x\right) - 1}$
   $\equiv \dfrac{\sin^2 x}{-\sin^2 x} \equiv -1 \equiv$ RHS

£1 million   *D is the correct answer.*

Section Three — Logs and Exponentials

1)a) $3^3 = 27$ so $\log_3 27 = 3$

 b) *to get fractions you need negative powers*
  $3^{-3} = {}^1/_{27}$
  $\log_3 ({}^1/_{27}) = -3$

 c) *logs are subtracted so divide*
  $\log_3 18 – \log_3 2 = \log_3 (18 \div 2)$
  $= \log_3 9$
  $= 2$ $(3^2 = 9)$

2)a) *logs are added so you multiply — remember* $2 \log 5 = \log 5^2$
  $\log 3 + 2 \log 5 = \log (3 \times 5^2)$
  $= \log 75$

 b) *logs are subtracted so you divide and the power half means square root*
  $\frac{1}{2} \log 36 – \log 3 = \log (36^{\frac{1}{2}} \div 3)$
  $= \log (6 \div 3)$
  $= \log 2$

3) *This only looks tricky because of the algebra, just remember the laws:* $\log_b (\chi^2 – 1) – \log_b (\chi – 1) = \log_b \{(\chi^2 – 1)/(\chi – 1)\}$
 Then use the difference of two squares:
 $(\chi^2 – 1) = (\chi – 1)(\chi + 1)$ *and cancel to get*
 $\log_b (\chi^2 – 1) – \log_b (\chi – 1) = \log_b (\chi + 1)$

4)a) *Filling in the answers is just a case of using the calculator*

x	-3	-2	-1	0	1	2	3
y	0.0156	0.0625	0.25	1	4	16	64

Check that it agrees with what we know about the graphs. It goes through the common point (0,1), and it follows the standard shape.

Answers

b) *Then you just need to draw the graph, and use a scale that's just right.*

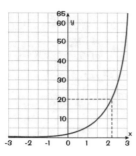

c) *The question tells you to use the graph to get your answer, so you'll need to include the construction lines, but check the answer with the calculator.*

$x = \log 20 / \log 4 = 2.16$, but you can't justify this accuracy if your graph's not up to it, so 2.2 is a good estimate.

5)a) $x = \log_{10} 240 / \log_{10} 10 = \log_{10} 240 = 2.380$

b) $x = 10^{2.6} = 398.1$

c) $2x + 1 = \log_{10} 1500 = 3.176$, so $2x = 2.176$, so $x = 1.088$

d) $(x - 1) \log 4 = \log 200$, so $x - 1 = \log 200 / \log 4 = 3.822$, so $x = 4.822$

6) *First solve for $1.5^P = 1,000,000$*

$P \times \log_{10} 1.5 = \log_{10} 1,000,000$,
so $P = (\log_{10} 1,000,000) / (\log_{10} 1.5) = 34.07$.
We need the next biggest integer, so this will be $P = 35$.

Section Four — Sequences and Series

1)a) $a = 2, r = -3$
10th *term*, $u_{10} = ar^9$
$= 2 \times (-3)^9 = -39366$

b) $S_{10} = \dfrac{2(1-(-3)^{10})}{1-(-3)} = \dfrac{1-(-3)^{10}}{2} = -29,524$

2)a) $a = 2, r = 4$, so $S_{12} = \dfrac{2(1-4^{12})}{1-4} = -11,184,810$

b) $a = 30, r = \frac{1}{2}$, so $S_{12} = \dfrac{30\left(1-\left[\frac{1}{2}\right]^{12}\right)}{1-\frac{1}{2}} = 59.985 \text{ (to 3 d.p.)}$

3)a) $r = 2$, so series is divergent

b) $r = 1/3$, so series is convergent

c) $r = 1/3$, so series is convergent

d) $r = 1/4$, so series is convergent

4) a) $r = 2^{nd}$ *term* $\div$ 1st *term*
$r = 12 \div 24 = \frac{1}{2}$

b) 7th *term* $= ar^6$
$= 24 \times (\frac{1}{2})^6$
$= 0.375 \quad (\text{or } \frac{3}{8})$

c) $S_{10} = \dfrac{24\left(1-\left[\frac{1}{2}\right]^{10}\right)}{1-\frac{1}{2}} = 47.953 \text{ (to 3 d.p.)}$

d) $S_\infty = \dfrac{a}{1-r} = \dfrac{24}{1-\frac{1}{2}} = 48$

5) $a = 2, r = 3$
You need $ar^{n-1} = 1458$, i.e. $2 \times 3^{n-1} = 1458$, i.e. $3^{n-1} = 729$. You can use then logs (or trial and error) to find that $n - 1 = 6$, i.e. $\underline{n = 7}$.

6) 1 5 10 10 5 1

7) $1 + 12x + 66x^2 + 220x^3$

8) $29120x^4$

9) $(2 + 3x)^5 = 2^5(1 + \frac{3}{2}x)^5$

$= 2^5[1 + \frac{5}{1}\left(\frac{3}{2}x\right) + \frac{5 \times 4}{1 \times 2}\left(\frac{3}{2}x\right)^2 + ...]$

x^2 *term is* $2^5 \times \dfrac{5 \times 4}{1 \times 2}\left(\dfrac{3}{2}\right)^2 x^2$,

so coefficient is $2^5 \times \dfrac{5 \times 4}{1 \times 2} \times \dfrac{3^2}{2^2} = 720$

Section Five — Differentiation

1)a) *A point where* $\dfrac{dy}{dx} = 0$.

b) $y = x^3 - 6x^2 - 63x + 21 \Rightarrow \dfrac{dy}{dx} = 3x^2 - 12x - 63$, *so set this equal to zero (and divide by 3) to get that the stationary points are where $x^2 - 4x - 21 = 0$, i.e. $(x - 7)(x + 3) = 0$, and so the stationary points are $(7, -371)$ and $(-3, 129)$.*

2) *Differentiate again to find* $\dfrac{d^2y}{dx^2}$.

If this is positive, stationary point is a minimum; if it's negative, stationary point is a maximum.

3) $\dfrac{dy}{dx} = 3x^2 - \dfrac{3}{x^2}$; *this is zero at $(1, 4)$ and $(-1, -4)$.*

$\dfrac{d^2y}{dx^2} = 6x + \dfrac{6}{x^3}$; *at $x = 1$ this is positive, so $(1, 4)$ is a minimum; at $x = -1$ this is negative, so $(-1, -4)$ is a maximum.*

4) a) $\dfrac{dy}{dx} = 12x - 6$, *so function is increasing when $x > 0.5$, and decreasing when $x < 0.5$.*

b) $\dfrac{dy}{dx} = -\dfrac{2}{x^3}$, *so function is increasing when $x < 0$, and decreasing when $x > 0$.*

Answers

Section Six — Integration

1) Check whether there are limits to integrate between. If there are, then it's a definite integral; if not, it's an indefinite integral.

2) The area between a curve and the x-axis, starting at the lower limit and going up to the upper limit.

3)a) $\int_0^1 \left(4x^3 + 3x^2 + 2x + 1\right)dx$

$= \left[x^4 + x^3 + x^2 + x\right]_0^1$

$= 4 - 0 = 4$

b) $\int_1^2 \left(\frac{8}{x^5} + \frac{3}{\sqrt{x}}\right)dx = \left[-\frac{2}{x^4} + 6\sqrt{x}\right]_1^2$

$= \left(-\frac{2}{16} + 6\sqrt{2}\right) - (-2 + 6) = -\frac{33}{8} + 6\sqrt{2}$

c) $\int_1^6 \frac{3}{x^2}dx = \left[-\frac{3}{x}\right]_1^6 = -\frac{1}{2} - (-3) = \frac{5}{2}$

4) a) $\int_{-3}^3 \left(9 - x^2\right)dx = \left[9x - \frac{x^3}{3}\right]_{-3}^3 = 18 - (-18) = 36$

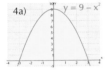

b) $\int_1^\infty \frac{3}{x^2}dx = \left[-\frac{3}{x}\right]_1^\infty = 0 - (-3) = 3$

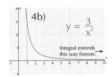

5) $\int_1^8 y\,dx = \int_1^8 x^{-\frac{1}{3}}\,dx = \left[\frac{3}{2}x^{\frac{2}{3}}\right]_1^8$

$= \left[\left(\frac{3}{2} \times 8^{\frac{2}{3}}\right) - \left(\frac{3}{2} \times 1^{\frac{2}{3}}\right)\right] = \left(\frac{3}{2} \times 4\right) - \left(\frac{3}{2} \times 1\right) = \frac{9}{2}$

6)a) $x_0 = 0$: $y_0 = \sqrt{9} = 3$

$x_1 = 1$: $y_1 = \sqrt{8} = 2.8284$

$x_2 = 2$: $y_2 = \sqrt{5} = 2.2361$

$x_3 = 3$: $y_3 = \sqrt{0} = 0$

$h = \frac{(3-0)}{3} = 1$

$\int_a^b y\,dx = \frac{1}{2}[(3+0) + 2(2.8284 + 2.2361)] = 6.5645 \approx 6.56$

b) $x_0 = 0.2$: $y_0 = 0.2^{0.04} = 0.93765$

$x_1 = 0.4$: $y_1 = 0.4^{0.16} = 0.86363$

$x_2 = 0.6$: $y_2 = 0.6^{0.36} = 0.83202$

$x_3 = 0.8$: $y_3 = 0.8^{0.64} = 0.86692$

$x_4 = 1$: $y_4 = 1^1 = 1$

$x_5 = 1.2$: $y_5 = 1.2^{1.44} = 1.30023$

$h = \frac{(1.2 - 0.2)}{5} = 0.2$

$\int_a^b y\,dx \approx \frac{0.2}{2}[(0.93765 + 1.30023) + 2(0.86363 + 0.83202 + 0.86692 + 1)]$

$= 0.1 \times 9.36302 \approx 0.9363$

7)a) $A = \int_0^2 \left(x^3 - 5x^2 + 6x\right)dx$

$= \left[\frac{x^4}{4} - \frac{5}{3}x^3 + 3x^2\right]_0^2 = \frac{8}{3}$

b) $A = \int_1^4 2\sqrt{x}\,dx = \left[\frac{4}{3}x^{\frac{3}{2}}\right]_1^4 = \frac{28}{3}$

c) $A = \int_0^2 2x^2dx + \int_2^6 (12 - 2x)\,dx$

$= \left[\frac{2}{3}x^3\right]_0^2 + \left[12x - x^2\right]_2^6$

$= \frac{16}{3} + 16 = \frac{64}{3}$

d) $A = \int_1^4 (x + 3)\,dx - \int_1^4 \left(x^2 - 4x + 7\right)dx$

$= \left[\frac{x^2}{2} + 3x\right]_1^4 - \left[\frac{x^3}{3} - 2x^2 + 7x\right]_1^4$

$= \frac{33}{2} - 12 = \frac{9}{2}$

Index